# UFO HEALINGS

*True Accounts of People Healed by Extraterrestrials*

by Preston E. Dennett

Artwork by Christine "Kesara" Dennett

Wild Flower Press
P. O. Box 190
Mill Spring, NC 28756

**Library of Congress Cataloging-in-Publication Data**
Dennett, Preston E., 1965-
UFO healings : true accounts of people healed by extraterrestrials / by Preston E. Dennett : art-work by Christina "Kesara" Dennett. p. cm.
Includes index.

ISBN 0-926524-33-X
1. Healing--Miscellanea. 2. New Age movement. 3. Unidentified flying objects. 4. Life on other planets. I. Title.
RZ999.D46 1996
001.9'42--dc20 96-31331
CIP

COVER ARTWORK: Christina "Kesara" Dennett
COVER DESIGN: Maynard Demmon

Printed in the United States of America.

Address all inquiries:
Wild Flower Press
P. O. Box 190
Mill Spring, NC 28756

Wild Flower Press has made the commitment to use 100% recycled paper whenever possible.

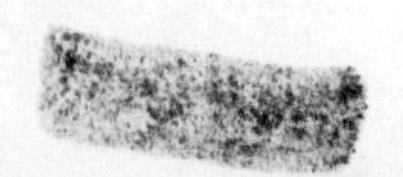

## Table of Contents

Introduction v

Preface ix

Why Aliens Make Good Doctors 1

Medical Evidence 9

Cures of Flesh Wounds and Other Injuries 17

Cures of Minor Illnesses and Ailments 41

Eye Doctors From Outer Space 55

Cures of the Integumentary System 65

Liver and Kidney Cures 73

Lung Cures 81

Healings of Serious Illnesses and Chronic Diseases 89

Cancer Cures 117

Experiencers and Psychic Healing 133

Other Miraculous Cures 139

A Chronology and Statistical Analysis 149

Epilogue 163

Biographical Sketch 171

Sources 173

Index 181

# Introduction

*by John Beresford, M.D.*

This is a remarkable book, in ways the author intended and probably did not intend. It is, in the first place, a step in the mysterious direction of healings associated with UFOs. What this means the reader must be left to decide. Healing itself implies "restoration of normal anatomy or physiology with or without the intervention of conventional medical techniques." The proverbial "shot of penicillin" leads to healing if an infective organism is "sensitive" to penicillin, the dose is right, and so forth. Essentially what happens is that the reproduction cycle of the organism is cut short, dead bugs are not replaced, and the body of the patient is left free to restore itself to normal. Surgical intervention accomplishes the same end.

"Spontaneous" healing is the term for statistically unlikely restoration of normal structure or function in the absence of deliberate medical intervention. Doctors are customarily said to be "baffled" or "amazed" when the statistically improbable healing occurs. What causes a return to normal when nothing has been done to invite it? Intervention by unseen beings, or angels, or beings higher on the scale of divinity, or human therapists endowed with an ability to move energy, or the kind of suggestion that is associated with placebo—the list of explanations is extensive. Spontaneous cases share the feature that conventional medical techniques have not been used.

UFO healing appears to depend on nothing like angelic intervention, suggestion, and so forth, but instead on medical techniques of an unearthly, advanced kind. We read of laser beams that slice through skin and muscle without damaging organs and blood vessels underneath. After some procedure, rent tissues are instantly mended, leaving, in the typical case, no evident trace of the foregoing intervention. Accounts are replete with descriptions of operating tables, beings dressed up like human doctors, "needles" for injection purposes (not otherwise described), and radiant flows of various kinds. The point is that the healing which occurs is almost magically fast—not magical, however, in that occult forces, as we like to think of them, are not responsible. Conceivably, after the intervention has been carried out; the body still conducts its own process of healing, but at an unimaginably speeded up rate. UFO healing incorporates medical conduct of some kind together with the rapidity of spontaneous healing.

This book makes an attempt to classify the rare cases of UFO healing reported in the UFO literature, with some cases traced by the author. Beyond a necessary association with spacecraft, the phenomenon itself is not defined. Extraterrestrial agents may or may not be definitely involved—a light beam alone may effect the cure. There may or may not be an abduction. Evanescent "hospitals" with staff may manifest, or, presumably, materialize, in a location where no hospital was formerly. The work of advanced medical intervention may be done in the subject's (one cannot very well say "patient's" bed at night. There appears to be no orthodoxy to the methods used to yield UFO healing. This makes the division of conditions along medical textbook lines not easy to follow. Few disorders act on a single organ, leaving the rest of the body unaffected. For example, a disease of the liver may lead to metabolic disturbance, with consequences in the heart and nervous system. UFO healing of a liver disorder, then, must involve more than structural and functional changes in the liver. Little of this comes through in the author's accounts of the episodes of healing he deals with.

"Give the man a break!" the reader will think. Dennett is not a medical doctor, and this book is not a medical manual. Nor was it meant to be. It is more an in-depth study and reporting of cases. Any lack of proper medical terminology in this book is, however, more than made up for by the fact that no professional medical doctor has had the background and nerve to write it.

The author has done what evidently no one else has done: single out for attention the beneficial effects that may result from UFO encounters ("UFO" standing for the range of UFO-related phenomena). People once sick have been rendered whole, according to these stories. In the wake of "kicking alien ass" in Independence Day, it is obvious that the most banal aspect of our culture—the movie is predicated on the glorious success of the Gulf War—is being polished up to reflect the view we are supposed to hold of aliens. Evil goes with slime. The ETs in Independence Day are a slimy, evil lot, bent on setting fire to the Earth's population centers. The movie is essentially a grade B cowboys and Indians saga, with ETs showing the loathsome qualities and in the end the weakness of Indians, and the President Clinton-like President and his sidelings the qualities of cowboy heroes. No room here for kind acts done to people.

The author makes out that UFO healings are commoner than reported. This may be so, but he has had to cast his net wide to catch the 105 cases he has managed to include. This brings up an unintended feature that makes the book remarkable. We are forced to think carefully of what constitutes evidence. Literary commentators on the first books of the Bible mention the skillful use of a multiple voice technique. God tells Moses something on a mountain where no one else is present. Moses tells the crowd at the foot of the mountain what God has told him to say. That is a double narration ("God told Moses who told me"). On what evidence is it to be believed that God really did say what Moses said he did? Then members of the crowd pass on the words communicated to them, eventually to the scribe who writes them down. That scribe's voice is the one carried in the narrative the reader reads. Again, on what grounds are events recounted in the final narrative deemed credible?

The same pattern appears in some of the included case reports. A woman goes by car on a trip. On the way an accident occurs. The hurt she receives is, however, healed by a UFO-related being. No one is there to verify the occurrence of either injury or healing when the woman arrives home. Do we believe her in the absence of collateral evidence?

On board a spacecraft an abductee is put through an examination and told of the existence of a medical disorder—which, however, will be put right. News of the presence of the disorder comes as a surprise; its presence was unknown. Following treatment, the abductee is returned to Earth. There he describes the diagnosis and the treatment, no trace of either now remaining. On what grounds is the evidence thought credible?

An alien heals a disorder that has been already medically diagnosed while a contactee is in the dream state. The experience remains vivid to the contactee, who wakes up finding the disability now gone. What is the evidence that an alien has been responsible for the healing?

Questions like these will pepper the thoughts of the careful reader, but many cases in this book present an astounding amount of evidence in support of UFO-related healings. So do go on with the book now and make up your own mind. The truth within these pages will speak clearly to your inner sense of knowing. Something important is evidently going on here.

# Preface

Anyone who has done even the smallest amount of research into the subject of UFOs knows that UFOs are real. An overwhelming amount of evidence supports this fact. The evidence comes in many forms including eyewitness testimonies, photographs, moving films, radar returns, metal fragments, landing traces, animal effects, medical effects, historical accounts and thousands of pages of documents from virtually every governmental institution in the United States. The question today is not if UFOs are real. There are much more important questions to answer such as, what are they? Why are they here? Where do they come from?

The most popular theory is that UFOs are extraterrestrial in origin, though there are a number of other lesser-known theories. UFO experts continue to argue over the question of why UFOs are here. The prevailing opinion is that they are neither invaders nor saviors, but are simply here to study humanity.

Dozens of books are published every year that present the vast array of UFO evidence. Because of the abundance of evidence, UFO books are now becoming specialized. There are books about abductions, books about the government cover-up, and books about UFO propulsion. The uninitiated skeptic is invariably overwhelmed by the huge number of different books published on the subject.

UFO skeptics and believers may both find the accounts in this book difficult to believe. I understand skepticism as I was a UFO skeptic for over twenty years. It was only a bizarre series of events that led me to begin a UFO investigation I never meant to follow.

It all began in 1986 when I heard an account on the evening news of a UFO sighting over Alaska. I didn't believe it for a second, but it interested me enough to ask my friends and family what they thought of all those ridiculous UFO stories.

Imagine my surprise when my brother said he saw a UFO, my sister-in-law said she had a face-to-face encounter with two alien entities, and two close friends said they were actually abducted! None of them had told me their encounters because they were embarrassed and didn't want to be ridiculed. They knew how cruel a skeptic could be, and they saw no reason to expose themselves to that kind of abuse. And so, like many, they kept silent.

I, however, had the opposite reaction. I instantly became obsessed with the subject and began reading every book on the subject. I quickly moved out of the armchair and into the field where I began interviewing witnesses and conducting on-the-spot investigations. In a matter of eight years, I interviewed nearly two hundred people, and uncovered UFO accounts of virtually every type.

There are many reasons why people remain UFO skeptics. Almost without exception, the UFO skeptic vastly underestimates the amount and quality of UFO evidence. Furthermore, prejudiced beliefs tend to blind people to the possibility of UFOs. But there is one main reason why so many people remain UFO skeptics. The reason is that UFOs are a package deal including the whole gamut of the unexplained. The UFO skeptic is instantly confronted with stories of levitation, telepathy, movement through solid objects, poltergeists, Bigfoot, and even worse...unexplained healings.

The typical reaction is to reject the entire subject as complete nonsense. The skeptic leaves the subject in disgust, horrified that people can actually believe such lurid accounts as say, alien abductions.

Skepticism, unless taken to extremes, is healthy. The problem, however, is that skeptics tend to ignore the evidence that doesn't fit into their world-view. The perfect example of this is the fact that when accounts of humanoids started appearing in the 1950s, many UFO investigators rejected the stories outright! It was only after hundreds of accounts were recorded that the UFO community began to accept the reports.

The same phenomenon occurred when abductions were reported. The idea that people could be taken inside a UFO and left with no memory of the event was simply too bizarre to be believed. However, as the accounts mounted, the evidence could not be ignored, and today, UFO abductions are the cutting edge of UFO research.

The accounts of UFO healings have often been relegated to the most unbelievable of UFO accounts. I remember my reaction when I first read of a UFO healing case. I just didn't believe it. My reasoning was simple: people who believe UFOs are here to cure us of our diseases must have a psycho-pathological need to believe in a higher power.

The idea that extraterrestrials are here to cure all our diseases is admittedly preposterous. If that were so, why would millions of people die every year from AIDS, cancer and heart disease?

But those pesky UFO healing stories wouldn't go away. They began popping up in books, articles, lectures and first-hand accounts of UFO witnesses. Having written over forty UFO articles myself, I decided to examine the accounts of UFO healings and see if there was any truth to the accounts.

I planned to write a small article, as I really didn't think that UFO healings were very common. To my surprise, I found over one hundred solid reports. It soon became obvious that if I wanted to include all the accounts, it would take a book. And so this book was born.

Doing the research for this book has been an awesome task, and it has radically changed my beliefs about UFOs. I discovered that there is a very fine line between an abductee (someone taken against their will aboard a UFO) and a contactee (someone invited aboard a UFO). After studying the accounts of healings, I began to look at UFOs more as floating hospitals than anything else.

The types of healings seemed straight out of science fiction. People reported their bodies being opened and closed with lasers that left no scars. They told how various organs were removed and put back again. They reported instantaneous cures of wounds and injuries. They reported healings of serious conditions like pneumonia or liver disease. They even reported healings of serious diseases, such as cancer.

I found strong parallels between the stories that seemed to exclude the possibility of hoaxes. In fact, the accounts were so consistent that it was obvious these people were telling the truth.

What follows are three typical scenarios of a UFO healing.

1. A person is awakened in his bedroom to see alien entities at the foot of his bed. The person is taken inside a UFO, given a physical examination, and told that he is sick with a disease. The aliens tell the abductee that they will perform a cure. After being probed with various instruments, the person is returned to the bedroom. Upon examination, all traces of the disease are gone.

2. A person is hospitalized because of injury, illness or disease. While alone in the hospital room, the patient is visited by a strange doctor who says she is there to help the patient. The "doctor" holds a small instrument over the patient and may administer medicine in the form of pills. The "doctor" leaves, often as mysteriously as she arrived. The patient quickly discovers that all symptoms of ill health have disappeared.

3. A person is driving along a road when a UFO makes a close pass over the car, sending down a beam of light. The person is engulfed in the beam of light. Suddenly, the UFO leaves and the driver is surprised to discover that she no longer suffers from an illness, disease or injury.

These accounts are reported by all types of people from all over the world. But because UFO healings are among the most incredible of all UFO stories, it is not too surprising that they are often ignored or given only brief mention.

Despite this overriding skepticism, many professional UFO researchers have realized the importance of such accounts. These researchers, although few in number, have made several positive statements concerning the veracity of UFO healings. If it was not for these brave pioneers, this book would never have been written.

Probably the best-known supporter of UFO healings is Edith Fiore, Ph.D., who has done extensive research into UFO abductions. As Fiore says, "One of the most interesting findings that emerged from this work was the many healings and attempts to heal on the part of the visitors.... In about one-half of the cases I've been involved there have been healings due to operations and/or treatments. Sometimes the cures are permanent. At other times the conditions recur.... If you have noticed a healing or inexplicable improvement...you may have had help from the visitors."[1]

David Jacobs, Ph.D., is one of the leading authorities on UFO abductions. Says Jacobs, "In extremely rare cases, the aliens will undertake a cure of some ailment troubling the abductee. This is not in any way related to the contactee/Space Brother concepts of benevolent aliens coming to Earth to cure cancer. Rather, in special circumstances, it appears that aliens feel obliged to preserve the specimen for

---

1. Fiore, 1989, pp. 322, 334.

their own purposes. As one abductee said, 'It's equipment maintenance.'"[2]

Budd Hopkins is probably the most famous UFO abduction researcher in the United States, if not the entire world. His two books, *Missing Time* and *Intruders,* have taken the UFO community by storm. At a UFO conference in Coronado, California, in 1994, Budd Hopkins admitted that his research has revealed UFO healing cases. As he says, "The question is whether we hear about healing cases. We do sometimes, very rarely, but they do turn up. And we don't know what to make of them. It's kind of a sad thing because I have some abductees who have serious medical problems who wish they were being healed themselves, but are not. So if they [the ETs] had the facilities, we wish they would do something. One of David Jacobs' clients said, 'I don't know whether I should be grateful as getting a present, or maybe it's just equipment maintenance.' So we really don't know. Incidentally again, there's no evidence whatsoever that this is a malevolent, evil conspiracy going on in the sky against us, they're going to take us over, or anything else. I'm very optimistic about the outcome because they seem to be most interested in what I consider the most human aspects of our personality, of our lives, of some of the most lovely aspects of being human. They're interested in that. There's no evidence, however, that they're here to help us. We wouldn't perhaps have AIDS and the hole in the ozone layer and everything else if they were here to help us. So there's no sense to this."

John E. Mack, M. D., has made quite a splash with his book, *Abduction: Human Encounters With Aliens.* Mack reports that several of his clients have experienced healings. As he says, "Some encounters are more sinister, traumatizing and mysterious. Others seem to bear a healing and educational intent...many abductees have experienced or witnessed healing conditions ranging from minor wounds to pneumonia, childhood leukemia, and even in one case reported to me first-hand, the overcoming of muscular atrophy in a leg related to poliomyelitis."[3]

Dan Wright is another UFO investigator who, as manager of the MUFON Abduction Transcription Project, is in a position to draw solid conclusions about the UFO phenomenon. According to the preliminary statistical analysis, 11% of physiological effects caused by

---

2. Jacobs, 1992, p. 191.
3. Mack, 1994, pp. 13, 45.

UFOs are healing cases. As Dan Wright says, "Almost one-third of the subjects reported some type of physical effect as a direct result of one abduction episode or another. Nose bleeds resulted in thirteen cases and scars in twelve, half of those on a leg or knee. Curiously, in four cases, the subject was either told by an entity or separately concluded that the beings 'reconstructive surgery' had repaired some medical problem."[4]

Leonard Stringfield is another well-known UFO investigator who has examined some of the accounts of UFO healings. As he says, "Healing cases on record baffle ufologists. More than a few who are looking into new realms for clues of the UFO nature and source are now seriously studying cases once dismissed as nonsense."[5]

Ralph and Judy Blum were among the pioneers of UFO research. They were also among the first to accept the reality of UFO healings. As Ralph Blum says, "Perhaps because they are as difficult to accept as contactee reports, reports of UFO-related healings are still scarce in the literature. And yet to me the possible connections between UFO light beams and paranormal healing is one of the most fascinating aspects of the phenomenon."[6]

UFO researcher Thomas E. Bullard made an invaluable contribution to ufology with his massive volume, *UFO Abductions: The Measure of a Mystery*. It represents one of the first objective studies of the UFO abduction complex. Out of 270 cases, Bullard reports that thirteen cases involved healings. As he says, "The other and more cheerful side of permanent aftereffects are the 13 instances where the witness left the abduction healed of some ailment.... Many of the cures appear to result from deliberate intervention, whereas the harmful effects could be accidental."[7]

Brad Steiger is one of the most popular UFO researchers and has written countless books on the subject. Because of this, he is well-aware of the UFO healing cases. As he says, "Many UFO percipients have reported miraculous healings, cures, even regeneration of teeth, after being touched by manifestations of UFO energy...over the years, several witnesses of UFO activity have reported rapid healing of cuts and the accelerated alleviation of certain illnesses after a close encounter with some aspect of the UFO experience. What is there

---

4. Wright, March 1994, pp. 5-6.
5. Stringfield, 1977, p. 72.
6. Blum, 1974, p. 143.
7. Bullard, pp. 149-150.

about contact with the UFO that can heal? Is it some electromagnetic radiation which might emanate from the object?"[8]

International UFO researcher Antonio Huneeus is quite aware that the energy from UFOs has had profound effects on people's bodies. He speculates that some of the physiological effects may be caused by an energy similar to microwaves. As he says, "It could well be that this microwave-like energy emitted by UFOs can heal under some circumstances, depending on such factors as intensity and proximity to the beam, and, of course, intent on the part of the UFO's occupants."[9]

Richard L. Thompson is a UFO researcher probably best known for his massive volume, *Alien Identities,* which draws parallels between accounts portrayed in ancient India's Vedic culture and modern-day UFO accounts. His book tackles many of the paranormal aspects of ufology, including healings. Says Thompson, "...There are reports of remarkable healings connected with UFO encounters. Some of these appear to be of a mystical nature. Others are attributed to medical interventions that seem to make use of recognizable high technology.... Of course, one can suggest that people imagine these ET cures because they need to explain natural cures occurring for unknown reasons. But Western culture provides familiar mystical explanations of unusual cures (such as the grace of Jesus). So why would someone try to explain mysterious cures by invoking even more mysterious ETs? The evidence that many UFO encounters tend to be accompanied by physical effects—injurious or beneficial—gives support to the hypothesis that these encounters are physically real. This is especially true in cases where the physical effect can be connected with recollections of specific events occurring within a UFO."[10]

The late D. Scott Rogo was one of the UFO investigators who was well-known for investigating the paranormal aspects of UFO encounters. Because of this, he is aware of UFO healing cases. As he says, "Writers and experts on healing—psychic, Christian Science, and others—often overlook one fascinating area of inquiry—those cases on record where psychic healings have occurred during UFO encounters!"[11]

---

8. Steiger, 1992, pp. 148-149.
9. Huneeus, Spring 1994, pp. 53-54.
10. Thompson, 1993, p. 128.
11. Rogo, 1977, p. 105.

Rogo, like most researchers, knows that there are a few famous cases, but as he says, "There are many other cases on record which report that UFO percipients have found themselves healed of all sorts of complaints.... In many cases, the healings seem to be linked to a mysterious light that was projected from the UFOs.... There are definite parallels between these cases. The appearance of the projected light beam, the nocturnal nature of the UFOs, and the unusually rapid healing of all follow a consistent pattern."[12]

Award-winning ufologist and editor of *UFO Universe* Timothy Green Beckley has this to say about UFO healing cases: "It is a documented well-established *fact* that UFOs have affected or been able to alter in some way the normal healing process.... Many miraculous healings have been—and are now—being reported, and in each case UFO activity is common; in some episodes, contact is made. No doubt we are dealing with an alien intelligence whose scientific methodology delves much deeper than our present-day technology can come close to duplicating."[13]

Jacques Vallée has written nearly a dozen books about UFOs. Regarding healing cases, he says, "We find that phenomena of precognition, telepathy, and even healing are not unusual among the reports, especially when they involve close-range observation of an object or direct exposure to its light."[14]

Another proponent of UFO healings is UFO investigator, Richard J. Boylan, Ph.D., author of *Close Extraterrestrial Encounters*. He has personally investigated healing cases of his own. According to Boylan, "Healing procedures are sometimes performed aboard UFOs. While the technology used is often so exotic that the human subject cannot tell what is being done or explain the equipment used, a number of humans have reported ET cures of conditions previously diagnosed by earth doctors as needing attention. In other instances, the ETs diagnose and cure the condition during the on-board experience. Treatments and cures have been reported for conditions like ovarian cysts, coronary valve disorder, vaginal yeast infection and obstructed nasal passage."[15]

Boylan reports that the examination given during a contact experience is much like a physical check-up. According to Boylan, "Some-

---

12. Rogo, 1977, p. 108.
13. Beckley, July 1988, pp. 6, 17.
14. Vallée, 1975, p. 21.
15. Boylan, 1994, p. 25.

times the results learned from the exam are communicated to the human, particularly if some worrisome condition is identified or if the human asks why a certain procedure was necessary. The purpose of the exam appears to be for the ETs to determine the subject's physical and genetic levels and their overall health status. Occasionally an ET will communicate to the human that a medical condition needs some attention and will indicate that either the human should consult an earth doctor about treating it or that it can and will be dealt with later. Sometimes, strong light, often colored, is shone down on the examinee, for what may be phototherapy."[16]

Boylan also knows that UFO occupants will actually make house calls. As he says, "On some visits, a medical scanning of a person is accomplished, apparently for diagnostic or follow-up purposes. Healing procedures have sometimes been conducted in bedrooms or other first-contact sites without removing the individual to a UFO."[17]

Although not extremely well-known, Paula Johnson wrote one of the first full-length articles about UFO healings. Says Johnson, "Documented cases are on record which show that UFO beings have the ability to cure our most deadly diseases and heal humans...there is ample proof that the pilots of these sparkling celestial craft we call UFOs *do* have the ability—and have on numerous occasions—cured humans of serious ailments, including cancer."[18]

Johnson is, of course, convinced that there are many more UFO healings than is publicly known. As she says, "...all this talk of aliens being able to cure cancer and other ailments isn't really that new. The files of various organizations are filled with similar accounts, adding weight to the notion that at least some ufonauts visit earth on healing missions.... How many UFOs are here on healing missions can't even be guessed at. It's possible that many who are being cured by aliens may not be aware of it. Others may be too timid to come forward to report their experiences."[19]

There are many other UFO researchers who are now uncovering these types of cases. One investigator I contacted, who wishes to remain anonymous, had this to say about UFO healings: "I am a certified hypnotherapist and I work with abductees seven days a week. I have many cases in my files where healing has occurred but people

---

16. Boylan, 1994, pp. 23-24.
17. Boylan, 1994, p. 22.
18. Johnson, July 1988, pp. 24-25.
19. Johnson, July 1988, pp. 26-27.

are afraid to tell their doctors what happened to bring about the healing."

The anonymous researcher is skeptical that aliens perform the healings out of pure kindness, and says, "I have a hard time dealing with space brothers who are here to help us. These beings have an agenda to follow and they are doing just that. They have not done one single healing out of the kindness of their hearts. It is for the end result that they do the healings. They have put forth a lot of effort on our people and they are not about to let us die before we have completed out mission for them. I have 160 cases that could tell you just the same thing I just said, but none of these people will come forward."

As we have seen, UFO healings may be rare, but they are not unique. They are a consistent feature of UFO encounters, one that can be ignored no longer. And although UFO investigators may disagree about the motive behind the healings, the accounts are finally being reported.

Unfortunately, there has never been a book about UFO healings...until now. This book represents the largest collection of UFO healing accounts ever assembled. It includes accounts from all over the world, many of which have never been published in book form, and some of which are brand new.

As any scientist knows, no data can be ignored without sacrificing the truth. Too many people have fallen into the trap of forcing the evidence to fit the hypothesis. Objectivity has always been the foundation of good science. Until all UFO accounts are examined, we will never have a complete understanding of the UFO phenomenon. To finish the puzzle, we need all the pieces. Only then will we have the whole picture.

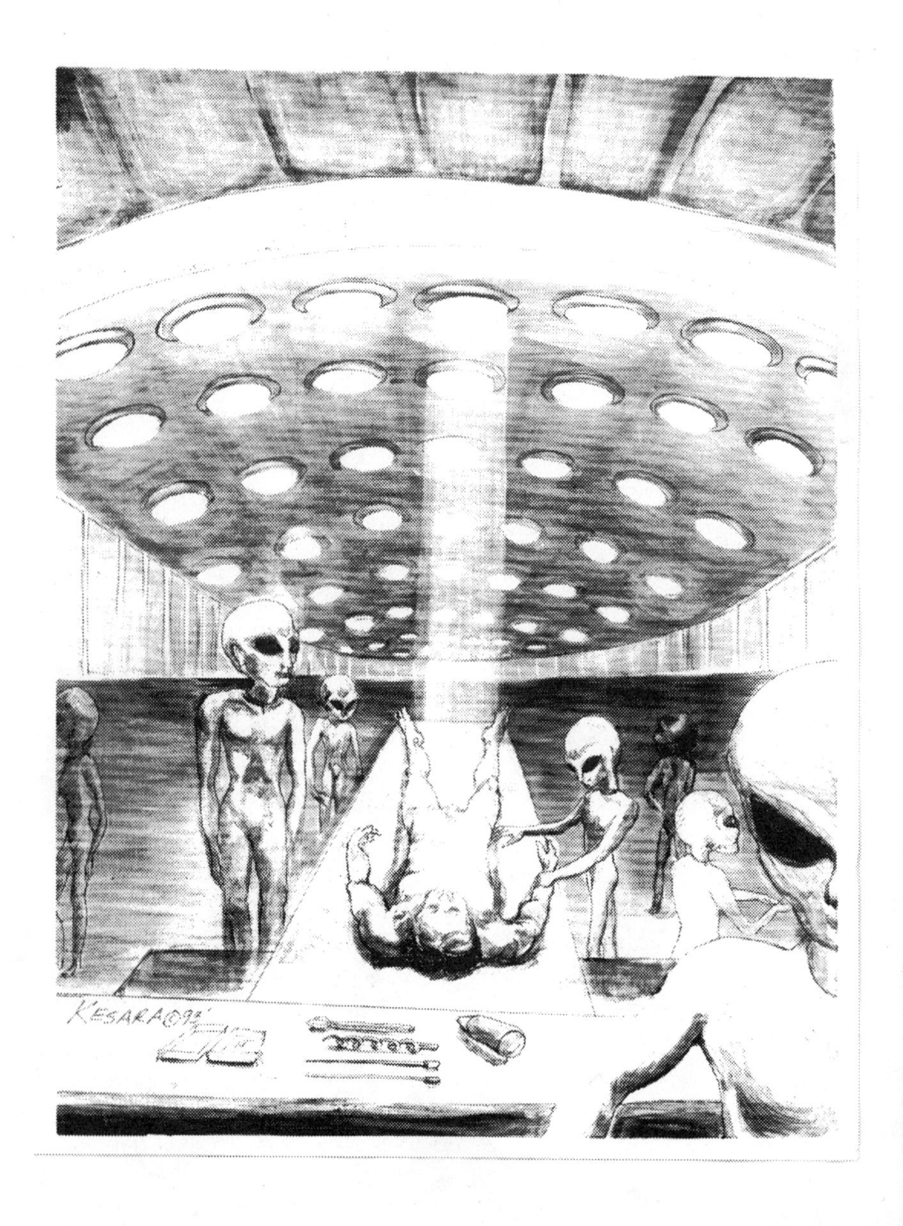
KESARA©

# Why Aliens Make Good Doctors

Since abductions became prevalent in the 1970s, they have become the focus of UFO research. It has now been recognized by dozens of prominent UFO researchers that abductions do occur. Unfortunately, some researchers have given the contact experience a negative interpretation. Words like "abduction" categorize the contact experience as a crime.

Unfortunately, the word contactee also carries negative connotations. And since there is no word for a person who has had a close encounter of the fourth kind, I shall use the term abductee as being synonymous with contactee.

Whether against their will or not, people are being taken inside UFOs, subjected to various procedures and then released. The number of abductions is recorded in the thousands, though some researchers estimate that millions of people may have had an onboard UFO experience.

Although there are many procedures performed upon humans aboard UFOs, the central and most consistent feature to all UFO abductions is the medical examination. In case after case, the details are the same: people report being paralyzed as they are laid out on an examination table. Samples of skin, hair, nails and reproductive material are often taken. The examination may involve lights or various instruments that are placed upon the body. In some cases, abductees report pain caused by needles or other instruments that are placed in virtually every orifice in the human body.

Other controversial reports are those of female abductees who report what has come to be known as the "missing fetus syndrome." There are scores of reported cases in which pregnant women have recalled losing babies after undergoing a gynecological procedure inside a UFO. With the aid of hypnosis, the witnesses recall that their babies were removed from their uterus by the aliens.

All of these cases have taught us many things. One fact, however, has become crystal clear. The aliens have a deep interest in the human body. The numerous types of scars that mark the bodies of the abductees illustrate this fact clearly.

What does this mean? One conclusion is obvious. Since the aliens are so interested in the human body, and since they have studied it so extensively, they know a great deal about how the human body works. This knowledge, combined with the fact that the aliens are equipped with very advanced technology, makes one fact very clear: aliens make very good doctors.

But is this true? Would aliens actually make good doctors?

All we really know about the aliens is what the people who have encountered them tell us. And in virtually every case, the abductee is first impressed by how closely the UFO room resembles a hospital room. Everything is clean and spotless. They are also impressed by the professionalism of their abductors. The examinations are typically brief, lasting about ten or fifteen minutes. Abductees all report that the aliens do their job with efficiency and speed.

Let us look at some of the cases to see if aliens really do make good doctors.

A well-known series of abductions occurred in 1953 in Tujunga, California. The story was told in a book called *The Tujunga Canyon Contacts*, and involves several women who were abducted aboard UFOs and given the standard examination. During Sara Shaw's examination, one of the aliens became interested in scars on her body. As Sara says, "They're looking at my lung surgery. My scar from my lung surgery…that scar fascinates them—they came over to look at that scar, too. They seem fascinated by that…"[1]

Sara actually used medical terms to describe the room. As she says, "They're taking me to an examining table. It's like an X-ray table."[2]

Sara was even told she was being examined. As she says, "They said—an electronic beam? Some kind of equipment, to examine me. That I don't have to be afraid of it, and it isn't going to hurt me."[3]

1. Druffel, 1980, 1988, pp. 26-27.
2. Druffel, 1980, 1988, p. 55.
3. Druffel, 1980, 1988, p. 55.

Another abductee, Sammy Desmond of Reseda, California, describes the inside of the UFO into which was taken. As he says, "It was a small room, all white, light colored and round. There was a table in the middle, and there were wall lights and things all over. The only thing I remember in the room besides the table was a huge, white, bright light in my face...it was a hospital-type skinny table."[4]

Desmond even describes his abductors' clothing using familiar medical terms. As he says, "They had white jacket-sort of outfits, sort of like hospital uniforms."[5]

Examinations aboard UFOs are remarkably similar to examinations in a doctor's office. Another abductee, Barbara X., describes her abduction: "They took a skin scraping...they looked at my bad leg...They said, 'You had an accident.' And I said, 'Yes.' They said, 'You had your female organs removed.' I said, 'Yes.'...they clipped my fingernails and a piece of my hair. They took a blood sample."[6]

Tom X., another UFO abductee, says, "I'm inside some sort of a roomlike area, on a table, undressed, nude, and I don't like this at all...they [the aliens] touched me with instruments. And they were cold, hard, probably metallic. They were prodding and probing and poking. Something was inserted in my penis and that was distinctly unpleasant, not painful."[7]

Abductee Bob Luca reports seeing the standard examination table aboard a UFO. As he says, "Looks like an operating table." He describes the examining instruments in similar terms. As he says, "Looks like a dentist's drill. It's folded up into the ceiling and there's a black thing on the end of it. I can't really draw it. But it looks like a dentist's drill. The arm comes out."[8]

Well-known abductee, Betty Andreasson has experienced many examinations at the hands of aliens. Needles have been inserted in her abdomen, her nose, her eyes and other places. As she describes one incident, "They're holding some needles by my head, and I just feel some things moving in there...they're putting something sharp in my heel so it feels like they're shooting something inside me."[9]

Another famous abductee, Barney Hill, describes the room he was taken to aboard a UFO. As he says, "I saw a hospital room. It was

---

4. Druffel, 1980, 1988, p. 310.
5. Druffel, 1980, 1988, p. 312.
6. Fiore, 1989, pp. 55-56.
7. Fiore, pp. 70, 73.
8. Fowler, 1994 Reprint, pp. 53, 55.
9. Fowler, 1994 Reprint, p. 152.

pale blue. Sky blue...it was spotless. I thought of everything being so clean."[10]

Barney's wife, Betty Hill, describes her abduction in similar terms. First she was put on a typical examining table, and then, as she says, "...the examiner opens my eyes, and looks in them with a light, and he opens my mouth, and he looks in my throat and my teeth and he looks in my ears, and he turned my head, and he looked in this ear...they take a couple of strands of my hair...they look at my feet, and they look at my hands, they look at my hands all over...and he cuts off a piece of my fingernail."[11]

Many female abductees report that their abdomens were pierced with a needle. Betty also reports this experience. As she says, "I asked the leader, I said, 'Why did they, why did they put that needle in my navel?' And he said it was a pregnancy test. I said, 'I don't know what they expected, but that was no pregnancy test here.'"[12]

At that time, in 1961, there were no pregnancy tests performed in that manner. However, today we have many similar medical procedures that are performed upon women, such as a lapyroscopy which allows doctors to peer inside the human body.

Budd Hopkins has done a great service in helping to bring UFO experiences under public scrutiny. Several of his first cases were recounted in his book, *Missing Time.*

Michael Bershad was one of the abductees featured in *Missing Time.* His account appears under the pseudonym of Steve Kilburn. Michael's physical examination involved the opening and probing of his back. Says Michael, "It's the one doctor's office I didn't have to wait to get into, I guess."[13]

Bershad spoke of his experiences in Los Angeles in 1992. During his lecture, Bershad spoke of his visit to Dr. Cooper, a professional neurosurgeon. Bershad explained how the aliens had stimulated certain nerves, causing certain reactions. Dr. Cooper was impressed that the aliens knew exactly what they were doing. As Cooper says, "He [Bershad] exactly described the motor reaction that happens when the femoral nerve is stimulated. And he has no particular knowledge of the nervous system...I'm really impressed with him. I told him that

---

10. Fuller, 1966, pp. 155-156.
11. Fuller, 1966, pp. 193-194.
12. Fuller, 1966, p. 196.
13. Hopkins, 1981, p. 80.

it seemed to me they [the aliens] just wanted to find out how he worked."[14]

Another abductee is Karen Morgan, whose story is told in Jacobs' book, *Secret Life*. Although she describes an unpleasant experience, the essential details remain consistent with other accounts. Says Karen, "I hate this room. There's tables...this room seems more like an operating room than any of the others...I think they're going to do that physical examination again. Oh, they take off my clothes...they turn me around, and lay me down. They pull off my jeans, my underpants. They strap me onto a table."[15]

The author of *Secret Life*, Jacobs has done extensive studies with abductees and has described the typical examination which is similar in many way to a conventional examination by human doctors. Jacobs reports that the aliens are interested in unusual marks on the body. As he says, "Scabs, infections, or other body marks and changes attract their curiosity. For instance, a woman who had given birth by cesarean section had new scars that drew the attention of the aliens who told her this was not the way *they* did it."[16]

Jacobs has investigated several cases involving highly technological medical equipment currently beyond our own capabilities. As he says, "The variety of machine examinations is great, although the exact purpose of the machines is unknown. Most abductees think they are recording devices, much like X-ray equipment. Somehow people know that the machines are scanning them, 'taking pictures,' or making neurological measurements."[17]

Travis Walton, who was abducted in Arizona in 1975, thought he was inside a hospital room when he was actually inside a UFO. He had just been struck by a blinding beam of light. His next memory was waking up in a hospital-like room. As he says, "They brought me into a hospital, I thought...I would let the doctors do all the worrying. I was safe for now.... Maybe I was in an emergency room of some kind."[18]

In 1986, I personally investigated an abduction case in which the witness, Kelly Robinson, was taken involuntarily into a UFO, laid on a table, examined, and told by the aliens that they were going to operate on her brain. As she says, "They tried to do something to my

14. Hopkins, 1981, pp. 87-88.
15. Jacobs, 1992, p. 89.
16. Jacobs, 1992, p. 94.
17. Jacobs, 1992, p. 133.
18. Walton, 1978, pp. 104-105.

brain and I was fighting them all the way." Subsequently, Kelly had three follow-up encounters in which she was given precognitive information about her personal life, her work and her church.

After having so many encounters, Kelly Robinson has come to know the aliens pretty well. She remembers exactly what happened to her and has never gone under hypnosis. Despite the fact that her experience terrified her, Kelly does not feel the aliens are hostile. As she says in her own words, "I think they're religious. They're not really out to hurt us. They're out to learn."[19]

Literally thousands of additional cases exist, and the examination is the one detail which varies least. Case after case involve these strange examinations. Although the cases differ in minor details, the pattern is almost always the same.

Many abductees are so conscious of the medical capabilities of the aliens, that they have given them the label of doctors. Abductee Shane Kurz says, "There is a table. Everything is so white...those eyes, they are telling me, lie down. They are taking my arm and scratching it. It hurts. And he is putting it on wax paper or something. It is squarelike and he gives it to the doctor. The doctor—I like him."[20]

Incidentally, Shane's abduction was terrifying and involuntary. However, Shane may have had the opportunity to become a contactee, and passed it by. One year before the experience, Shane was approached by a normal, yet odd-looking stranger, who invited her for a ride in his "white vehicle," which was allegedly parked, unseen, in a field. There were many strange details to the encounter that seemed to indicate that Shane was being invited aboard a UFO by an extraterrestrial. She declined the invitation.[21]

As we have seen, the descriptions of the cases corroborate each other to a remarkable degree. People who have no knowledge of other cases are all reporting the same details.

Despite all the controversy surrounding UFOs, there is one thing that most UFO investigators agree upon: UFOs represent a superior intelligence. UFOs have demonstrated time and time again that their machines are more powerful than our own. Their vehicles easily outdistance our fastest jets. They remove people from moving cars, dense suburbs, hotels and apartment buildings. They subject people to examinations and operations whose purposes we can only guess.

---

19. Personal files.
20. Holzer, 1976, p. 234.
21. Holzer, 1976, p. 222.

In addition, they often leave people with no conscious memory of what has happened to them.

The aliens have shown themselves to be outstanding pilots, superb hypnotists and excellent scientists. In the field of medicine, however the aliens are unparalleled. Their medical instruments and procedures are far in advance of our own. Their knowledge of the human body is also more extensive than our own. This is evidenced by not only the reports from the thousands of people who have been examined aboard UFOs, but from a collection of accounts of a much rarer type.

Most people have little physical evidence of their encounter other than perhaps a scar. Some, however, are given more dramatic proof of alien intervention. Often the proof amazes even experienced UFO investigators.

Because of this, accounts of UFO healings have been largely hidden from public view. However, these cases clearly show that aliens have the capability to cure virtually every illness or disease known to humankind. Although the cases are not well publicized, aliens have been actively curing people across the planet for over fifty years, possibly even longer.

KESARA ©
93'

# MEDICAL EVIDENCE

When the modern age of UFOs began in the late 1940s, investigators had access to only a limited amount of UFO evidence. Throughout the 1950s, UFO investigators were only able to record sightings and compare them with other sightings. In the 1960s, more evidence accumulated in the form of radar returns, photographs and films, landing traces and reports of alien creatures. Finally, in the 1970s, UFO abductions became widely publicized, and since then, the focus has shifted towards studying the accounts involving close interactions with UFO occupants.

As the number of reported onboard experiences increased, a new type of evidence began to exhibit itself. This type of evidence involved human physiological reactions to the close presence of a UFO. It soon became termed "medical evidence."

Medical evidence was quickly realized as being of profound importance, and before long, it became the pivotal detail compelling investigators to accept the truth of a UFO witness's claims. Medical evidence, it turned out, could make or break a case.

The Mutual UFO Network (MUFON), the world's largest civilian UFO research organization today, has chapters in every state in the United States, and over four-thousand members worldwide. MUFON field investigators are provided with an investigators manual and reporting forms for UFO witnesses to fill out.

One section of the manual focuses on medical effects. MUFON lists the most common physiological effects as: "tingle or shock, dizziness, noticeable body temperature change, unexpected joint or muscle stiffness, motor skills affected (e. g. spasms or paralysis), levitation of the witness, eye irritation or impairment of sight, ear irritation or impairment of hearing, nose irritation or impairment of smell, a burning sensation or actual skin burn, skin rash, cut or gouge, headache, loss of appetite, nausea, extreme fatigue, sleep disorder, or other."[1]

The MUFON form for physiological effects lists other possible symptoms including hair standing on end, hair burning, hair turning white, hair falling out, tooth fillings vibrating, vomiting, warts and skin peeling.

A few other effects worth mentioning that were not on the list include: menstrual disturbances, dehydration, bruises, scars and perhaps radiation sickness. In addition, the material fails to so much as mention UFO healings. In fact, field investigators are told that the only available medical evidence is destructive. The heading under the chapter is "Medical Injuries From UFO Close Encounters."

According to the manual, these injuries fall into three major categories. The first are those injuries of a temporary nature, the second are those of a chronic nature, the third are parapsychological manifestations. The manual instructs the investigator to obtain data on the witness's blood count, possible weight loss and urinary ketones, advising X rays as well as biopsies of any skin lesions.[2]

Obviously, the forms need to be updated. But as can be seen, the medical evidence for UFOs takes many forms and at first glance, the list is scary. The MUFON form also makes no mention of the few cases in which close encounters have resulted in the death of the witness, usually of symptoms that seem to parallel radiation sickness.

The veracity of a UFO case depends on many forms of evidence. The investigator is always happy to stumble upon a case with medical evidence, as such a case provides the smoking gun, the proof that the experience actually took place.

Thousands of cases on record involve medical effects. Let us examine some of the better-known cases to see more clearly the extent of the medical evidence for UFOs.

Possibly the most famous medical evidence case is the Cash-Landrum case which occurred in Texas in 1981. Three witnesses, Betty Cash, Vickie Landrum and Colby Landrum, all saw a UFO hovering in front of them on the road. The object was emitting heat waves and evidently other types of radiation. Immediately afterwards, all three began to suffer some alarming symptoms.

---

1. Fowler, 1983, p. 8.
2. Fowler, 1983, pp. 131-133.

Betty had a headache, neck pains, swollen and painful eyes, sunburn, nausea, vomiting, diarrhea, loss of appetite and severe hair loss. She was treated in Parkway General Hospital in Texas as a burn victim. Later she developed breast cancer. Vickie, who was further away than Betty, suffered eye irritation and minor hair loss. Colby, who was the farthest, suffered only a minor first-degree burn on his face and eye irritation.[3]

Another famous case is the Falcon Lake, Canada case, in which a prospector, Stephen Michalak walked right up next to a landed UFO. The object took off unexpectedly, blasting Michalak with heat and setting his shirt on fire. Because of the intense pain, and because he had a sudden headache and was vomiting, Michalak went straight to the hospital.

Soon after his arrival at the hospital, Michalak began to suffer a number of alarming symptoms including a headache, diarrhea, weakness, dizziness, nausea, vomiting, hives, numbness and swelling of joints, swelling of hands, a burning sensation around neck and chest, eye irritation, a peculiar body odor and fainting. He was unable to keep any food down and lost twenty-two pounds. His blood lymphocyte level decreased from a normal twenty-five to an alarming sixteen percent. Michalak was examined by over twenty-seven doctors, and the only explanation that seemed to fit was exposure to radiation.[4]

Yet another well-known case of medical evidence was the abduction of three ladies, Louise Smith, Mona Stafford and Elaine Thomas, in South Dakota. All three of the ladies saw a UFO while driving and experienced a period of missing time. Immediately afterwards, they suffered severe eye irritation, first-degree burns on their necks, fatigue, weight loss, nausea, vomiting, diarrhea. Under hypnosis, they were able to recall being examined by humanoid ETs.[5]

Still another classic case is the abduction of Antonio Villas Boas in 1957. His abduction involved many of the standard elements and also a sexual encounter with an exotic-looking alien female. After his abduction, Boas suffered from insomnia, nausea, headaches, loss of appetite, burned and watery eyes, bruises and scars. He was put under the treatment of a doctor who was unable to account for the symptoms.[6]

---

3. Good, 1988, pp. 303-304.
4. Good, 1988, pp. 197-198.
5. Lorenzen, 1977, pp. 114-131.

One interesting case involves a man who was hospitalized because of suspected kidney tumor. The doctors made this diagnosis due to "peculiar marks" found on his abdomen. The suspected tumor was never found, perhaps because the patient neglected to mention that the marks had appeared after an apparent UFO abduction.[7]

Hopkins' book, *Missing Time,* profiles the experiences of Virginia Horton. Immediately after one of her abductions, Virginia returned with a large cut, of which she still bears the scar. As she says, "I think my leg was cut with a scalpel. It was just really sharp and clean...as if somebody made a nice, clean, quick incision."[8]

On another occasion, Virginia was filmed immediately after an abduction. The film clearly shows fresh blood on her blouse. The blood had come from a puncture-wound caused by a nasal probe during her abduction. Incidentally, the aliens told Virginia that they had abducted her because, as Virginia says, "They were celebrating and they said they wanted to share it with me because my research or their research was—interesting."[9]

Abductee Filiberto Cardenas experienced many symptoms after his abduction in Hiahleah, Florida. His symptoms include insomnia, excessive thirst, excessive sexual appetite, changes in body temperature, peculiar body odor, skin abrasions and marks, headaches and eye irritation. Physicians at Jackson Memorial Hospital in Florida verified these symptoms.[10]

Another well documented case involving medical evidence is that of Denise Bishop, who saw a UFO outside her home in Plymouth, England, in 1981. As she watched the object, a beam came out and struck her hand, with a paralyzing effect. As she says, "As soon as it hit my hand I couldn't move. I was stopped dead in my tracks."

A few seconds later, the beam retracted and Denise rushed back inside. Shortly later, she noticed that the beam had burned her hand. As she says, "On looking at it I noticed spots of blood and after washing it saw it was a burn." A few days later, the burn seemed to become worse, so Denise saw a doctor who said that the burn was typical of lasers. Denise still bears the scar where she was burned by the UFO.[11]

---

6. Bowen, 1969, pp. 234-236.
7. Hopkins, 1981, p. 22.
8. Hopkins, 1981, p. 138.
9. Hopkins, 1981, pp. 201-204.
10. Sanchez-Ocejo, 1982, pp. 40-46.
11. Good, 1988, pp. 98-99.

In 1964, eight-year-old Charles Keith Davis of New Mexico experienced even more extensive injuries as the result of a UFO encounter. Charles was outside playing in his backyard when an object swooped down out of the sky and covered him in a "blackish ball of fire." He was rushed to the hospital and treated for severe burns.[12]

Equally impressive is the case of James Flynn of Florida. While hiking out in the Everglades, Flynn encountered a UFO that flashed him with a beam of light. Flynn immediately lost consciousness. When he awoke, the UFO was gone, leaving him partially blind in his left eye, and totally blind in his right eye. In addition, his forehead and face became red and swollen. Flynn eventually recovered, except for a partial cloudiness in his right eye.[13]

One man was even able to receive worker's compensation after his injurious encounter with a UFO. Nightwatchman, Harry Sturdevant of New Jersey was "buzzed" by a UFO. Shortly afterwards, he began suffering from nausea and loss of his sense of smell and taste. He was unable to swallow and soon collapsed in pain. Six weeks later, a worker's compensation referee ruled in Sturdevant's favor, requiring that he be reimbursed for all medical expenses caused by his UFO encounter.[14]

In South America, on the islands around Belem, there are dozens of cases involving people who have been struck by UFO beams that cause a number of alarming symptoms. Dr. Carvalho, the director of the community health center of Marajo, has verified all these symptoms. The symptoms include weakness, dizziness, headaches, low blood pressure, anemia, burns, puncture wounds and hair loss. In several of these cases, the witnesses died shortly after being struck by the UFO beam.[15]

Probably the most well-known of deaths caused by UFOs is that of Joao Prestes Filho of Brazil. Prestes was struck by a beam of light from a UFO. Shortly afterwards, his flesh began to fall from his bones as if boiled. He was rushed to the hospital, but died before he arrived.[16]

Well-known UFO investigator Richard Hall wrote an outstanding UFO reference book called *Uninvited Guests*. The book outlines over thirty separate cases involving physiological effects. Hall pre-

---

12. Steiger, 1967, p. 15.
13. Steiger, 1967, pp. 21-22.
14. Steiger, 1967, pp. 22-23.
15. Vallée,1990, pp. 130-139, 217-223.
16. Vallée,1990, pp. 124-125.

sents virtually the entire range of medical effects, including a few of the well-known healing cases.[17]

A startling number of UFO cases contain some degree of medical evidence. Nearly every abductee can offer one form or another of medical evidence, such as a scar appearing in conjunction with a UFO encounter.

Another common medical complaint is that of eye irritation. The least common is death of which there are few reported cases. Significantly, the majority of the injury cases appear to be accidental. The scars caused by medical procedures aboard UFOs are, of course, not accidental, but even here on Earth, surgery leaves scars. Many of us bear the tell-tale marks of vaccinations.

Despite the number of reports of UFO encounters with medical effects, the vast majority of UFO encounters involve no medical effects at all. Rarely is anyone injured by a UFO encounter. In fact, healings may be just as common as injuries, if not more so.

The central issue, however, is that the medical evidence of UFO encounters plays a crucial role in determining the credibility of a case. Medical evidence is particularly important, because it is often the only hard physical evidence available to the UFO researcher. Virtually all UFO investigators agree that medical evidence represents a vital piece of the UFO puzzle. For this reason, we must investigate all available medical evidence.

Early accounts of UFO healings were slow to be recognized. At first, the accounts were not believed, and many received very little publicity. Over time, however, more and more witnesses began reporting miraculous cures. Gradually, UFO investigators became increasingly interested in their stories.

Today, however, UFO healings are becoming much more well-known. New accounts come out every month. And as we shall see, the cures performed range from minor ailments to life-threatening diseases.

17. Hall, 1988, pp. 231-232.

3

# Cures of Flesh Wounds and Other Injuries

Many of us have experienced the pain of cutting ourselves with a knife or getting into an accident and sustaining a number of painful bruises. Sometimes these injuries can be serious enough to warrant an immediate visit to the doctor. Unfortunately, a doctor can only treat a flesh wound, not cure it. Our medical science has not advanced to the level of Star Trek technology, where a single laser can make a cut simply disappear. The only known cure for a flesh wound is time. The human body repairs itself, but at a slow rate. Therefore, if we hurt ourselves, we often suffer for days, weeks or months afterwards.

But this is not always the case. On several occasions, people have been cured of their flesh wounds by UFO encounters. These ET cures have also been effective on cases of broken bones and even paralysis.

Of all the UFO healings, one of the most often cited is the healing of a flesh wound. This case has been recounted in several UFO books, and remains a classic un-disputed case in mainstream ufology. What was healed in this case was a cut on a policeman's finger caused by his son's pet alligato.

CASE #017. On September 3, 1965, two police officers, Patrol Deputy Robert W. Goode and Chief Deputy Billy McCoy, were driving along Highway 36 South in Damon, Texas. The two officers were returning from a high school football game.

Deputy Goode was suffering from a painful although minor injury. He had been bitten on the left index finger by his son's pet baby alligator. The finger was red and swollen.

Suddenly, Chief Deputy McCoy saw strange lights rising from the right side of the road. McCoy pulled the patrol car off the road and pointed out the lights to Deputy Goode. When Goode looked, he too saw the lights.

At this point, the lights moved towards the police officers. They weren't able to see any detail until the lights were very close. Then they saw a solid mass at least two to three hundred feet long and fifty feet wide. There was a bright purple light on the left side and a blue light on the right. Goode estimated that the object was about the size of a football field as he could see the huge shadow the object cast on the ground.

Suddenly, the object moved over the patrol car and cast down a beam of light. As McCoy says, "The inside of the car was lit up by the bright light."

Goode's left arm with his injured finger was hanging outside of the window when the beam struck. As Goode says, "I could feel the heat from the light. We got out of there."

The two terrified officers raced away at speeds approaching one hundred miles per hour. They kept going until they were back in the city, where they pulled into a diner to discuss the incident.

It was then that Goode noticed that his finger was no longer throbbing with pain. As he says, "I suddenly realized it was not bothering me and I pulled it out of the bandage. Hell, you couldn't tell I had ever been bit."

McCoy verified Goode's story. As McCoy says, "The swelling had disappeared and the finger looked a lot better."

The two deputies decided to return to the location of the incident to see if the object was still there. To their surprise, it was. When the UFO started to move towards the police car, the two officers sped away and refused to return.

The sighting was reported to the Air Force, and Major L. R. Leach of Ellington Air Force Base conducted an investigation. The results of the investigation were that the object remains unidentified. Later, the two officers were besieged by newsmen and their story was told across the country. Today, the account can be found in many UFO books.[1]

In the above case, the healing was obviously caused by the beam of light from the UFO. There are several other cases on record involving people who were struck by a beam of light and reported cures of flesh wounds. The following well-known case has also been well-sub-

---

1. Green, 1967, pp. 85-86.

stantiated. It is also very rare in that the witness was cured of both a flesh wound and partial paralysis.

CASE #021. On November 2, 1968 in the French Alps, "Doctor X" was awoken just before 4:00 A.M. by the cries of his infant son. He found his son looking out the window, gesturing towards something. Looking out through the shutters, the doctor saw strange flashes of light, which he first thought were lightning.

A few minutes later, he looked out a clear window and saw two luminous disks hovering outside his home. They were white on top and red on the bottom, and flared brightly at regular intervals. As he watched, the two disks moved to the left, converged and became one. The single object then shone a beam of light down upon the ground. After a moment, the beam swept forward and struck the doctor on the front of his body. There was the sound of a loud bang and the object disappeared.

At the time of the incident, Doctor X suffered from a partial paralysis of his right arm and leg. The paralysis was caused by injuries received many years earlier, during the Algerian war. Before the war, Doctor X was able to play the piano quite well. After the war, the paralysis rendered him unable to play the piano at all, and he also walked with a limp. Doctor X also suffered from a recent flesh wound on his ankle. He had been chopping wood three days earlier when the axe slipped and gouged deeply into his ankle, which was still swollen and painful.

After being struck by the beam of light, the doctor was dazed and shaken. He awoke his wife and told her what happened. She instantly noticed that her husband's ankle wound had "completely disappeared." The next day, Doctor X realized that he no longer suffered from any type of paralysis. He no longer walks with a limp and has taken up piano playing again.[2]

A similar case of the healing of a flesh wound comes from UFO investigator Timothy Green Beckley. In this case, the witness was healed during what appeared to be a simple sighting. As it turned out, there were other strange effects as well.

CASE #034. On July 8, 1974, at 10:15 P.M., New York City syndicated columnist, Brandon Blackman reported seeing a UFO over

---

2. Vallée, 1975, pp. 21-24.

Prospect Park in Brooklyn, New York. As Brandon says, "I saw this glow wobbling in a falling leaf motion. Stopping at a particular level, it would hover and rise again several times."

Several other witnesses saw the object, including Blackman's fiancee. The object approached the witnesses quite closely and then departed.

When Blackman returned home, he discovered that his house keys were missing. He knew he had the keys with him when he left because he double locked the door. The keys, he thought, must have been lost outside. However, once inside, the keys were found lying in the center of the bed. Blackman feels that the UFO may be responsible.

Blackman also reports that the day of the UFO sighting, he had accidentally slashed his finger. He quickly bandaged it and forgot about it. However, as he prepared for sleep, he removed the bandage and was very surprised at what he found. As he says, "When I removed the wrapping, I discovered much to my amazement that the wound was completely healed, as though absolutely nothing at all had happened!"[3]

Some cases of UFO healings are shrouded in mystery, not because the healing is unexplained, but because the illness is. As many people have discovered, doctors are not always able to correctly diagnose a condition, and the patient must live with the illness or condition until their doctor can prescribe a treatment. Unless, of course, the patient is healed in another way, as in the following case.

CASE #014. On September 27, 1959, in Coos Bay, Oregon, real-estate salesman, Leo Bartsch was cured of an unexplained numbness in his left arm and fingers. Bartsch had been suffering from the condition for three weeks and his doctor was unable to find any cause for his numbness. As Bartsch says, "The trouble extended up my arm. It felt dead, as if blood did not want to enter it."

Then, in the middle of the night, Bartsch was woken up with a strange feeling of weightlessness and electricity. He had the distinct impression that "something out of the universe had just passed over...."

---

3. Beckley, July 1988, pp. 15-16.

Immediately afterwards, Bartsch discovered to his delight that all numbness was gone and he could move and control his fingers again. Bartsch believes he was cured by a UFO.[4]

Contactees appear to experience a great number of UFO healings. Being involved with extraterrestrials evidently has its benefits. Miraculous healings are only one of the many wonderful experiences that contactees are privileged to have. In many of these cases, the contactee is actually taken inside of the UFO and cured of whatever illness demands attention. The following case is a perfect example.

CASE #033. In January of 1974, contactee Dr. Frank E. Stranges drove out of Las Vegas, Nevada in order to rendezvous with the extraterrestrials out in the desert. Stranges claims to be in contact with friendly extraterrestrials who look like normal humans. The aliens claim to come from Venus, and Stranges' alien friend is named Valiant Thor.

On the way to meet Valiant Thor, Stranges was attacked along a desert road by several men in a large black car. The men physically assaulted Stranges. As he says, "I was thrown to the ground and kicked over and over again." Before long, Stranges had sustained numerous bruises and a badly crushed hand.

In the middle of the assault, Valiant Thor showed up and rescued Stranges from the attackers. Stranges was taken inside a UFO where he was treated for his injuries. As Stranges says, "As I lay atop the soft white table, I was conscious of the beam of soft blue light that emitted from the cone-shaped instrument which was aimed first at my head, then at my solar plexus. I immediately went into a deep sleep. Upon awakening, I felt good all over."[5]

Contactees are often given a difficult mission by their extraterrestrial friends. These missions typically include taking steps to help end wars, poverty, pollution and hunger. Contactees are also often expected to tell about their experiences in hopes of raising the spiritual awareness of humanity. Because of these missions, it is essential that the contactees maintain good health. However, sometimes accidents occur. And occasionally, these accidents seem to warrant alien intervention, as is true in the following case.

---

4. Johnson, July 1988, pp. 26-27.
5. Stranges, 1981, pp. 43-47.

CASE #027. In 1971, in southern California, contactee, Dr. Fred Bell and two of his friends were the victims of a crude pipe-bomb explosion. All three were injured, however, Bell was thrown off his feet and onto the sidewalk where he lost consciousness. He was rushed to Laguna Memorial Hospital. When he awoke, he discovered that his left hand, upper left shoulder and neck were badly burned. He also sustained a minor skull fracture and many painful bruises all over his body.

Dr. Bell claims to have been in contact with human-looking aliens from the Pleiades. The contacts began as a child and continued throughout his life. They have given him life-saving warnings and valuable advice. They also gave him an innocuous-looking necklace called a "receptor." Bell was told that it would protect him from harm.

While alone in the hospital room, Bell felt the necklace heat up and send "marvelous energy" into his body. Within minutes, he was completely healed. His visiting friends watched with disbelief as he removed the bandages and declared himself cured. The next morning, Bell had the enjoyable experience of watching the hospital physicians run a number of tests in an attempt to explain his miraculous cure.

As one of the physicians reportedly said, "This is impossible. You have no trace of the severe burns that had damaged your left hand, upper left shoulder, and your neck. There is no longer evidence of a skull fracture. And what with all those bruises, eh? You were covered with bruises. Where did they disappear to?"

The doctor was reportedly very upset and actually accused Bell of having faked the injuries. However, because Bell appeared to be in perfect health, they were forced to let him go. Bell did not tell the doctors about the "receptor" necklace, which he believes cured him with some "unknown energy."[6]

Another similar case of a UFO healing occurred to a contactee who also claims to be in contact with the Pleiadians. This case has never been published before and was investigated by the author.

CASE #040. Sometime in the mid-1970s, contactee Anthony Champlain (pseudonym), of Florida, was taken aboard a UFO and healed of an injury. Anthony claims to be in contact with extraterres-

---

6. Bell, 1991, pp. 78, 88-89, 118-119.

trials from the Pleiades. Like Fred Bell and other Pleiadian contactees, he has experienced many adventures aboard the alien ships.

Unfortunately, Anthony doesn't recall the exact date of his healing, nor does he recall exactly what type of injury he had. He admits that his memory of the event is somewhat sketchy. Yet he knows that the injury was on his arm and hand, where he had hurt himself by falling. He also remembers being taken aboard the Pleiadian craft where he was healed. His good friend was also taken aboard and examined.

Like many people, Anthony was cured of his injury by a beam of light. As he says in his own words, "It was a slow-moving beam of light. It was a slow-moving beam, and it slowly hit you. And you sort of watched it. She said, 'It isn't going to hurt.' Because I thought, 'Is it going to hurt?' She said, 'No, it isn't going to hurt.' I said, out loud, 'What are you doing?' She said, 'Is there some part of your body that is broken?' I said, 'Yes, well you're fixing it?' She said, 'Yes, we're fixing it.' Not fixing, repairing, something like that.

"To this day, "Anthony continues, "I don't remember what it was that they fixed. Because I kept trying to think, 'What was broken?' I can't remember what was broken. That was the funny thing about it. I remember some discussion that something was broken. I remember that they were repairing it. And I don't remember what it was that was broken."

There may actually be a good reason why Anthony can't remember. As is well-known in the UFO field, some ETs often block out people's memories of their time aboard the ship. As Anthony says, "I think it was blacked out. I wasn't sure what it was. I remember, I got up. What I do remember is that, when I was laying down, I was no longer in clothes. I remember some aspect of it. She said, 'Are you ashamed?' I said, 'No.' Something like that. I forget the actual details of it, but my other friend was, and he was watching, and he was thinking, 'This guy's naked!' And she asked him, 'Are you ashamed?' And he said, 'Yes.' So, I don't know. He went into another room."[7]

Many people have endured the extremely painful experience of bumping their heads. If bumped hard enough, a huge lump will be raised, which often takes weeks to heal. Head injuries are also extremely dangerous because of the possibility of brain damage. Our ability to deal with head injuries is often limited to reducing pressure

7. Personal files.

in the skull with drugs and other techniques, as well as locating damaged areas of the brain with various sensing instruments. Other than radical brain surgery, there is little the doctor can do to treat a head injury.

In the following case, however, extraterrestrials were able to cure a head injury using remarkable methods.

CASE #044. James X. is today a prominent physician in Southern California. As a young child, sometime in the early 1970s, he was playing on the jungle-gym when he fell off and hit the back of his head. A huge goose-egg sized lump appeared and his father took him to the hospital that night to have the lump drained. After that incident, he suffered severe headaches.

One day, the headaches simply stopped. Unknown to James at the time, he had been abducted and cured aboard a UFO. It was only later, in his mid-thirties that James underwent hypnosis and discovered his remarkable experience.

He recalled being aboard a UFO and undergoing a terrifying examination. At the time, he had no idea why the abduction was occurring. He remembers being strapped to a table where a strange helmet was placed over his head. As he says, "...there's something on my head, like a hemisphere over the top part of my head, and it feels like it's clamped on over my maxilla."

James watched with curious fascination as the aliens dissolved the lump on his head using what looked like laser beams. As he says, "They go in there and dissolve that damn clot...there's two beams, and when they cross, it creates heat at that exact location. And it takes a long time, because it dissolves the clot, gasses form, and they don't want to obstruct the blood flow."

James, as a physician, recognized that the aliens were merely speeding up a process that would have occurred naturally. Nevertheless, the cure is beyond the scope of today's medical expertise.[8]

Another case involving the healing of a head injury comes from a gentleman who changed his legal name to Star Traveler, presumably as a result of his many encounters with grey-type aliens he calls the Zetas.

---

8. Fiore, 1989, pp. 166-197.

CASE #090. "Star Traveler" reports that his UFO experiences have been "informative, pleasant and even funny..." One of these experiences includes the healing of a head injury. As a young boy, Star Traveler was playing softball when his sister accidently struck him in the head with a baseball bat. As he says, "The bat struck my skull just above my eyebrow and the flesh peeled back like the lid of a sardine can."

In the emergency room, the doctor applied a butterfly bandage, and advised that he might need stitches. The next evening, Star was in his room lying in bed when he saw a bright star-like object outside his window. In the middle of the night he woke up to see two aliens with "chalk-white skin and large, black almond-shaped eyes that seemed like peaceful, glistening pools of water. They were skinny, and had large heads, thin necks and very long arms."

As he lay there, one of the figures began passing a rod-shaped instrument back and forth over his body. At first he was frightened, but then the aliens told him what they were doing. As he says, "I realized they were there to heal me, and I couldn't apologize enough! They told me (telepathically) that they routinely heal 'volunteers' (those who volunteer to be of service to others before they incarnate) and insure their health. It was their job to 'make the rounds.'"

Needless to say, Star Traveler enjoys meeting with the ETs. As he says, "I consider them my friends."

A few days after the head injury, Star Traveler visited the doctor again. As Star says, "The physician was shocked to see that the injury was healing at an amazing rate—and that I would need no stitches."

Star Traveler says that he has been healed on many different occasions. As he says, "They have effected many other healings for me, including healings of an impact injury to my feet and pleurisy in my lungs. I have come to regard these interventions as a 'Zeta-Terran medical plan' that comes as a bonus for being a 'Volunteer.'"[9]

Not all reports of healings involve beams and instruments. Some cures seem to be effected by psychic means. For example, in the following case a lady was cured of a flesh wound by the prolonged stare of an extraterrestrial person.

CASE #015. Sometime in the 1950s, contactee Ann Grevler of the Eastern Transvaal in South Africa, encountered a landed UFO. She

---

9. Traveler, 1995, pp. 9-10.

was subsequently invited aboard the ship and was given a tour through space.

However, prior to her invitation, the aliens showed her firsthand just how advanced they were. Before her very eyes, the aliens made Grevler's car invisible, using a "rod-like device." Believing that her car was gone and not just invisible, Grevler walked over to where her car should have been. As she did, she "gashed her leg on the invisible license plate."

The alien told her that although they could turn objects invisible using mind power, it was much easier with the use of gadgets. The undescribed spaceman then cured Grevler using mind power alone. As the report says, "...the spaceman was able to administer first aid, accomplished by directing his gaze to the wound, which promptly healed."[10]

Another case which has never been published comes from a southern California MUFON field investigator and medical doctor. Although he remains skeptical that the case in question is actually a UFO healing, he does admit that a miraculous healing took place, and that the person healed has been abducted on numerous occasions. And although the case does have unique elements, in most respects it fits the typical pattern of a UFO healing.

CASE #079. An anonymous southern California family has experienced countless sightings, dozens of close-up encounters, and several abductions. All members of the family have had experiences, and the contacts are continuing.

The family was driving through a remote desert area in the south western United States. It was a very hot day and the car over-heated. The father pulled off to the side of the road and opened the hood. He then made a terrible though common mistake. While the car engine was still hot, the father opened the radiator cap.

Instantly, searing hot liquid exploded out of the hole, badly burning the father's hand. The family rushed to his aid, wrapping up the father's hand. They quickly put more water in the radiator and drove for the nearest town. On the way there, the family was very surprised to encounter what appeared to be a hospital in the middle of the desert.

---

10. Good, 1993b, p. 75.

The man got out of the car and walked into the strangely located "hospital." Inside, there was a nurse who said she would help him. She held his wrist and sprayed a jelly-like substance over his hand. The substance formed a layer over his hand, covering the burn. The "nurse" then instructed him not to wash his hand, and he would be fine.

By the time the family made it into town, the father's hand was completely cured. The family then realized that something strange had happened, and that the hospital was probably not what it seemed to be. They returned to the scene and, not surprisingly, they were unable to locate the "hospital" in which the father had been cured.

The MUFON investigator who handled this case was baffled by the appearance of the "hospital." He could not understand how aliens could convince people they were in a hospital and not a UFO.[11]

Actually, there are similar cases on record. One example comes from Budd Hopkins' book, *Intruders*, in which the main witness, Kathie Davis (Debbie Jordan), experienced many abductions. After one abduction, her only memory of the event was visiting a strange store, which she thought was the local 7-11!

In addition, there are many reports of strange and advanced instruments being used to cure various wounds. These instruments differ in size and shape, but are often hand-sized instruments that emit some type of light or heat. In the following unique case, an unknown type of laser beam was used to repair an incision that had been made by the aliens themselves.

CASE #056. In October of 1988, Fred X. of New York City, was abducted out of a friend's apartment. His memory of the event was virtually non-existent. The only indications that something had happened were a time lapse in his memory and several scratches, marks and bruises on his body. Later on he began having some recall of the event in his dreams; he eventually underwent hypnosis and recovered complete memory of the event.

When his recall was complete, Fred was able to recount the entire incident. Fred remembers being taken aboard a craft. He was given a standard medical examination, and sperm was removed. While in the medical room, he saw other operations occurring. He observed one

---

11. Personal files.

animal injected with a needle which extracted fluid from the animal's body.

Fred also saw a nude woman lying perfectly still on a table. To Fred's horror, he saw that the woman had a large surgical incision down the center of her chest. As Fred says, "She's been opened up and has a vertical incision from the top of her chest down to the groin area."

Unable to move, Fred watched the grey-type aliens insert their hands inside the woman's chest opening. Fred then noticed that the woman's legs were spread-eagled with clamps around her ankles, and a "long tube-like instrument" was inserted in her vagina.

Fred had the impression that the examination of the woman, himself and the animal were part of a related experiment involving reproduction. As Fred says, "This was strictly surgical."

At one point, Fred saw another grey-type alien approach the woman with a hand-held object with a light on it. The alien pointed a laser-like beam of light at the incision and healed the cut. As Fred says, "What he is doing to the skin, as he pulls it together, it's just sealing it up as if there wasn't any cut.... He uses the light, pulls the skin together, and you can't tell she was ever cut."[12]

This is not the only account of a healing of a wound that was caused aboard the UFO itself. Abductees and contactees spend many hours aboard UFOs. The possibility always exists that an unforeseen accident could occur aboard these UFOs. Evidently, just this event happened to a contactee who slipped and fell while aboard the UFO, as this next case illustrates.

CASE #062. On October 15, 1989, contactees Bert and Denise Twiggs of Hubbard, Oregon, woke up in their bedroom to discover many strange marks on their bodies. Because of previous extensive experiences with aliens who claimed to be from Andromeda, Bert and Denise knew that the marks meant that they had experienced a contact during the night.

Bert noticed that his back was very sore, and he wondered what had transpired during the night to cause his back pain. Both Bert and Denise decided to undergo hypnosis to recover their memories. Both easily remembered what happened.

---

12. Turner, 1992, pp. 161-169.

While aboard the UFO, Bert had tripped and fallen backwards against a metal support bar, injuring his back. The two of them were returned to their home, and later, the extraterrestrials made a house-call to heal Bert's back. As Denise says, "The doctor placed a hand-held machine over the sore area. The machine reversed most of the damage that occurred when Bert hit his back on the metal bar. After this process was finished, they left...and Bert awoke nearly healed."

It is important to note here that the Twiggses claim to have had a number of illness cured by their Androme friends. These cases are covered in later chapters.[13]

The Twiggses may have collectively received more healings than any other contactees. In the following case, Denise Twiggs had just delivered her baby by Caesarean section. After the surgery, she was visited in her hospital room by her alien friends who came to assist her recovery.

CASE #048. In November of 1981, Denise Twiggs gave birth to her son, Christopher by Caesarean section. She had planned to deliver Christopher naturally, but unforeseen complications demanded immediate surgery. It seems that the umbilical cord was wrapped around Christopher's neck.

Luckily, there were no difficulties, and Christopher was born a healthy, normal baby.

Denise was disappointed to have a C-section as the recovery process is longer than with natural birth. However, Denise had alien friends who actually visited the hospital room and sped up the healing process of her incision. As Denise says, "...the Androme doctors visited me, giving me a helping hand with my recovery. My human doctor was amazed at how quickly I was healing."

The human doctor, of course, had no idea about Denise's encounters.[14]

Of course, it is very rare for a person to receive a cure from extraterrestrials. Some people, however, report numerous beneficial experiences aboard UFOs. Almost without exception, those who claim to have repeated healings aboard UFOs are contactees. Bert and Denise Twiggs of the above cases are contactees who have experienced sev-

---

13. Twiggs, 1995, pp. 111-112.
14. Twiggs, 1995, p. 98.

eral healings. The following case involves a man who was cured of multiple injuries and illnesses as they occurred throughout his life.

CASE #006. One winter in the early 1950s, in New Jersey, Richard Rylka, then a young child, was playing outside in the snow with his friends. They were sleigh-riding down steep hills near their homes. While racing down the hill, Rylka was unable to steer and he crashed into a car that was driving up the hill. The driver didn't even see Rylka who was pushed under the car, thrown off the sled and knocked unconscious. By the time the driver found him, Rylka had been laying on his left side in the snow for many minutes. His left ear was badly frostbitten and he developed an instant ear infection which destroyed his hearing in that ear. His doctors considered performing a mastoidectomy, but instead they decided to try a radical new drug called penicillin.

Rylka, although a child, had seen UFOs on several occasions. He remembers going outside and being beamed messages from the aliens. As of yet, however, they had not visited him physically. As he lay in the hospital, Rylka was surprised when the ETs, named Koran and Nepos, came to the hospital itself for the sole purpose of curing him. The aliens were able to partially cure Rylka. Using "their own body energies," the aliens were able to get rid of the pain and pressure. After the aliens left, Rylka was still unable to hear. But following several more doses of penicillin, his hearing returned.[15]

CASE #024. Sometime in the 1960s, while a young man (Rylka was born in 1941), Rylka was working for a medical firm. On one particular occasion, he was instructed by his supervisor to transport a barrel of chemicals. To do this, Rylka had to operate a piece of heavy equipment called a "drum-lifter." Heavy machinery is often dangerous, and somehow Rylka's index finger became caught in the machinery and crushed. Rylka's screams called the attention of co-workers who helped him to free his finger, which was "white and terribly distorted." He soon lost feeling in the finger and was unable to move it.

Rylka was rushed to the company's first aid room and the nurse told him to go to the hospital at once. Rylka took a cab to Middlesex General Hospital in Brunswick, New Jersey where doctors applied an ice-pack and prepared for X rays. Rylka was alone in the emergency

---

15. Randazzo, 1993, Chapter 11.

room when the ETs, Koran and Nepos, appeared, and held their hands over this finger and cured him using body energies. Although the finger was numb for many months, the cure was effective enough to straighten the bone and close the broken skin. When the doctors returned, they found no signs of injury and sent him back to work.

Later, Rylka was contacted by the company nurse who had read his hospital file with disbelief. She told Rylka flat-out, "You were supposed to lose that finger." She demanded to know what happened, but Rylka remained silent about his alien friends.[16]

CASE #026. In 1970, Rylka had a devastating automobile accident in New Jersey. His body was so badly mangled, he had a near-death experience. While many people see deceased relatives during a near-death experience, Rylka saw his alien friends who told him that his body was so badly injured, he should be dead. They promised to heal him, and also warned him of drastic changes to his life at home and work.

All the predictions of the aliens came true. Rylka also received other cures which will be covered in later chapters.[17]

A very similar case of a healing after an auto-accident took place in Atlanta, Georgia. I was able to interview the main witness in depth regarding the incident.

CASE #053. Alicia Hansen (not her real name) has been having contacts for many years. In 1987, in Atlanta, Georgia, she miraculously survived what everyone thought should have been a fatal accident. As it turned out, the accident was fatal and Alicia had been literally brought back to life by her alien contacts.

As Alicia says, "I was in a really bad car accident, very bad. I got hit head on. I was a passenger in the car. I knocked out the side window with my face and then I went face-first through the windshield. I had to go to the emergency room and everything. They thought my neck and back were broken. I went face first through the windshield...I had to go through physical therapy for a year, learning how to walk again. I had a lot of internal damage done. My jaw was dislocated and all kinds of other stuff."

Despite her extensive injuries, the paramedics couldn't understand why Alicia hadn't been hurt more badly. What was strange was

---

16. Randazzo, 1993, Chapter 11.
17. Randazzo, 1993, Chapter 11.

that her face was completely unmarked. Alicia had no memory of being healed. All she remembers is waking up to find herself lying in the street. It was years later, under hypnosis, that Alicia found out the truth of what really happened that night. As Alicia says, "I was doing a regression one night and I saw that I had actually died in that accident. I didn't have a cut, I didn't have a scratch on my face. I knocked out two windows, went through the windshield, and I had not a scratch. And everybody thought that was really weird...but I saw that I broke my neck, and I saw that they were doing surgery on me. They were fixing my brain stem. They were putting me back together. I saw all of that. And my face was just devastated. I saw this brain stem had been broken, damaged, and they repaired that. And my face was really cut and they just whoosh!—they fixed it. It was the weirdest thing I've ever seen. It was just healed. And I didn't have a cut, not a scratch, nothing. They left certain things, but they fixed the most important. And they were also doing something. They said my metabolic rate was really messed up."

Alicia doesn't understand how the aliens were able to heal her before the paramedics even arrived, however, she learned from later experiences that the aliens have the ability to travel in time.

During the experience, Alicia also says that there was "an exchange of souls to some degree." She says, "I'm kind of hesitant to talk about it because I don't understand it completely yet...but the weird thing about it was when I saw this, and I saw this accident, there was some sort of an exchange. Like a higher part of myself changed places with myself."[18]

Swiss farmer Eduard Meier is without a doubt one of the most famous contactees of all times. He claims to have had hundreds of contacts with human-like aliens from the Pleiades. Meier's story is supported by a large amount of evidence including moving films, photos, audio recordings, landing traces, metal fragments, animal evidence, medical evidence and multiple eye-witness testimonies. Nevertheless, the case is extremely controversial.

Meier, like most contactees, has been instructed by his alien friends to tell others of his contacts. For this reason Meier has reported his experiences publicly. Like most contactees, Meier, as we shall

---

18. Personal files.

see, has received a number of healings. The following case is a typical example of an injury cured by extraterrestrials.

CASE #037. On April 3, 1976, contactee Eduard Meier of Switzerland was instructed by the aliens to go to a certain location where a flight demonstration of the alien crafts would take place. Meier was told that he could bring his friends and also take photographs.

Unfortunately, the group was followed by unwanted human visitors who caused the demonstration to be canceled. Meier was instructed telepathically by the aliens to depart the area. As he was leaving, he was pursued by the unwelcome observers. In his haste to escape, he ended up crashing his moped. The crash damaged the motorbike and also hurt Meier who had a fractured rib, a dislocated shoulder, a twisted foot and several abrasions.

Three days later, Meier was visited by his Pleiadian friend, Semjase, who offered to cure his painful broken rib. She held a small device to the broken rib, and instructed Meier to sit between the two "poles" on the instrument. She told Meier, "The rib-bones will be completely regenerated after this process. There will be nothing left to indicate that they had been broken."

Meier felt a sudden "electrical sensation" and then nothing. He knew instantly he was cured because the pain was gone. Semjase was unable to reduce the swelling only because the correct instrument was not available at the time.

Meier's injuries and cure were verified by other witnesses. Incidentally, Meier has only one arm, having lost one in a bus accident. During one of his contacts, the aliens offered to grow him a new limb. Meier declined, fearing that people would ask too many questions.[19]

As we have seen, the healing of a flesh wound by UFOs is usually done by beams of light. But these healings are not always limited to humans. There is at least one case on record in which a human saw an alien cured aboard a UFO.

CASE #012. In 1957 and 1958, Cynthia Appleton of Aston, Birmingham, in West Midlands, England, had four separate contacts with an alien entity. The alien was six feet tall, had a pale complexion, long blond hair and elongated facial features. The alien communicated with Appleton using telepathy. He claimed to be an extraterrestri-

---

19. Stevens, 1990, pp. 251-258.

al and said that open contact with humanity was impossible because of wars. The alien gave Appleton a large amount of highly technological information. Physical evidence was also left in the form of scorch marks on a newspaper on which the alien stood.

During one of the contacts in February 1958, the alien accidentally burned his finger. He asked Appleton to bathe his finger in hot water, which she did. The alien then injected himself with a tube and sprayed a jelly-like substance on his hand. According to the report, the wound "healed miraculously."

There is physical evidence supporting this event. After the healing, Appleton found a piece of the alien's skin in the bowl of water. It was analyzed by professionals under an electron microscope at Manchester University.

The investigator in charge of the case, Jenny Randles, was able to speak with the doctor who examined the skin. The doctor showed Randles his report which showed that the skin was not human skin, but seemed more similar in composition to animal skin. The doctor also admitted that he was unable to discover from which animal the skin came. The doctor has declined to go public with his report.

During her contacts, the aliens also gave Appleton a method of curing cancer. This will be presented in a later chapter.[20]

Extraterrestrials heal people in many differnet ways. In the following case, a woman was given the power to heal herself. Shortly following a very intense UFO encounter, the witness was able to heal herself of a serious flesh wound and an unexplained condition. The case also illustrates how some UFO encounters seem to involve reproduction.

CASE #072. At 4:31 A.M., on January 19, 1992, Northridge, California, was struck by a 6.8 earthquake. However, the night before the earthquake, Topanga Canyon resident, Michelline (not her real name) experiences a dramatic UFO sighting over her home. She and her boyfriend saw a large diamond-shaped craft towing a smaller craft using a beam of light as a tow-rope. The craft hovered over nearby water towers before moving slowly away. A few hours later, the earthquake struck.

---

20. Randles, 1988, pp. 70-72.

Two weeks later, Michelline had another visitation. However, on this occasion, she saw a grey-type alien standing next to her in her bedroom. She tried to get up but discovered she was paralyzed. Oddly, she felt no fear. As she says, "It kind of looked like it was a cat face without any hair and a combination of a skeleton face.... I tried getting up, and my legs would not move. After a little while, I actually had an orgasm, just sitting on the couch. The weird part was, while I was sitting there looking at this thing, I was trying to write down the things it was telling me. It was telling me that they were trying to invent ways of intercourse. They needed babies. They were just telling me a bunch of things telepathically; there would be more earthquakes. And then you know we had all those earthquakes all over the world.

"A few weeks later, I all of a sudden had this bulge in the left side of my stomach. I had gone to the doctor, and I went to the emergency room. They wanted to do surgery. They said, 'You know, maybe it could be a hernia,' or 'We're not really sure what it is.' I didn't want to have an operation. And I decided in my head, this is going to go away. And it went away three days later.

Three days after that, I dropped a seventy-five pound slate of marble on my foot. It didn't cut my foot, but you could see where there was an indentation in my foot, probably being broken. Two days later, that went away. And then about four months later, I had a miscarriage and there was nothing in the sac."

Michelline credits her sudden healing abilities to the extraterrestrials. As she says, "When I saw them, the thing that I wrote down was that I would be able to heal myself. And I would be able to heal people. If they wanted me to heal them, I would be able to heal them."[21]

John Mack reports that in one of his cases, a young girl was comforted by ETs after being sexually molested. In the following case, investigated by southern California therapist and UFO researcher Barbara Lamb, M.S., M.F.C.C., a young girl was healed of her injuries resulting from a molestation.

---

21. Personal files.

CASE #007. This case involves a young four-year-old girl from southern California. The case occurred in the early 1950s. The details were uncovered through hypnotic regression, conducted by Barbara Lamb. As she says, "She had an experience at age four where she was sexually abused by her grandfather, and there was actually some physical damage done to her in her genital area because of that. And fortunately the act was stopped sort of in the middle by the grandmother who discovered them. Anyway, later on that afternoon, probably an hour or so later, this lady was taken by her 'little people,' the little gray extraterrestrials, and taken aboard a spacecraft. And they did a sort of whole medical healing procedure on her, the little gray guys did. And then there was a female [alien] too, who was a little bit larger. And the female was standing by her side on the medical table on the craft. And the female was explaining that something was done here that should not have been done, and that the other ones, the little ones were needed to fix this, to repair this, because they wanted to be sure that everything would be fine in later years for having babies."[22]

Going to the dentist is something few people enjoy. As least one person may have avoided such a visit by getting a little help from another type of dentist. This case was investigated by Bob Teets, former journalist and author of the book, *West Virginia UFOs*. In this case, a man was cured of pain caused by impacted wisdom teeth.

CASE #076. This case involves a man by the name of Daniel D. from Harrison County, West Virginia. Daniel reports that he has had many, many contacts with "gray guys" and "tall green, lotus-looking figures." He definitely believes the aliens are friendly. As he says. "They've never hurt me, that means a lot...I think they've helped me in certain ways. I've heard for years that I have to have my wisdom teeth out. One day they were bothering me really bad, and I was in no financial situation to get them fixed, no insurance. That night I was abducted. I can remember, it felt like they were stuffing things on my teeth where my wisdom teeth are. The next day they didn't bother me anymore and haven't bothered me since. This was a couple of months ago."

Daniel D. reports that during his abduction, he actually asked the aliens telepathically if they would fix his teeth, and they speedily obliged.[23]

---

22. Personal files.

Although not common, some animals have been healed by UFOs. The following case may well be the only case on record in which healings have been extended to animals.

CASE #013. One summer afternoon, in 1959, in Pleasanton, Texas, Susan Nevarez Morton, who was thirteen years old at the time, went to attend a cockfight with her sister and brother-in-law, and their family. The entire group drove to the secret location and joined a huge crowd of people on wooden bleachers. In the center of the crowd, two roosters began to fight a battle which Susan knew would be to the death.

The two roosters were fiercely fighting when suddenly, the attention of the crowd was caught by a large, glowing object which was hovering over the area. Suddenly, the object emitted two beams of light. Susan described the object as "an enormous red globe with two shafts of white light inching down towards us."

Everyone watched in amazement as the beams targeted the two roosters, who were now lying still on the ground, nearly dead from their injuries. Then something amazing happened. As Susan says, "Their small broken bodies glowed eerily for a few seconds. Then slowly, they both got up on their little chicken feet and began strutting around with robust healthy enthusiasm."

The crowd became extremely agitated, until, after a few moments, the beam of light retracted inside the object. Then the object changed from red to orange, and streaked away at high speeds. The two roosters showed no trace of their former injuries, but the cockfight was canceled.[24]

As we have seen, aliens have no difficulties curing injuries. The cures being performed across the world. Yet another case of a UFO healing occurred to a young boy from Russia.

CASE #093. A boy from Tblisi, Georgia in Russia injured his knee. While still suffering from the injury, the boy received a visit by friendly extraterrestrials who "operated" on his knee, restoring it to normal.[25]

As these cases illustrate, extraterrestrials are able to cure a huge variety of injuries. The healings were usually performed with beams

---

23. Teets, 1995, pp. 26-27.
24. Morton, Feb. 26, 1989.
25. Morrow, Jan./Feb. 1993, p. 17.

of light that resemble lasers. Some report heat, while others report no feeling at all. The beams are described as very bright, and either blue, purple or white in color. They are either emitted from a hand-held instrument or from the UFO itself. The lights are typically described as being unlike normal beams. As one witness said, the beam was extended and retracted very slowly.

While a few of the cases seem to be accidental, the vast majority were obviously planned events

It is interesting that none of these people claim to be cured of every injury they have ever sustained. Often, the witnesses are perplexed as to why one particular injury was healed and another was not. If the aliens have the ability to cure all injuries, why don't they? And why were these people cured of their injuries when other far more influential people are often injured and/or killed with no intervention from extraterrestrials?

Unfortunately, until we have more data, these questions must remain unanswered. The data shows only that the cures are happening to people of all ages, races and backgrounds. There seems to be no discernible pattern as to who is cured.

Not too surprisingly, injuries are not the only cures that extraterrestrials perform. They are also quite proficient at dealing with other minor illnesses and ailments of all types.

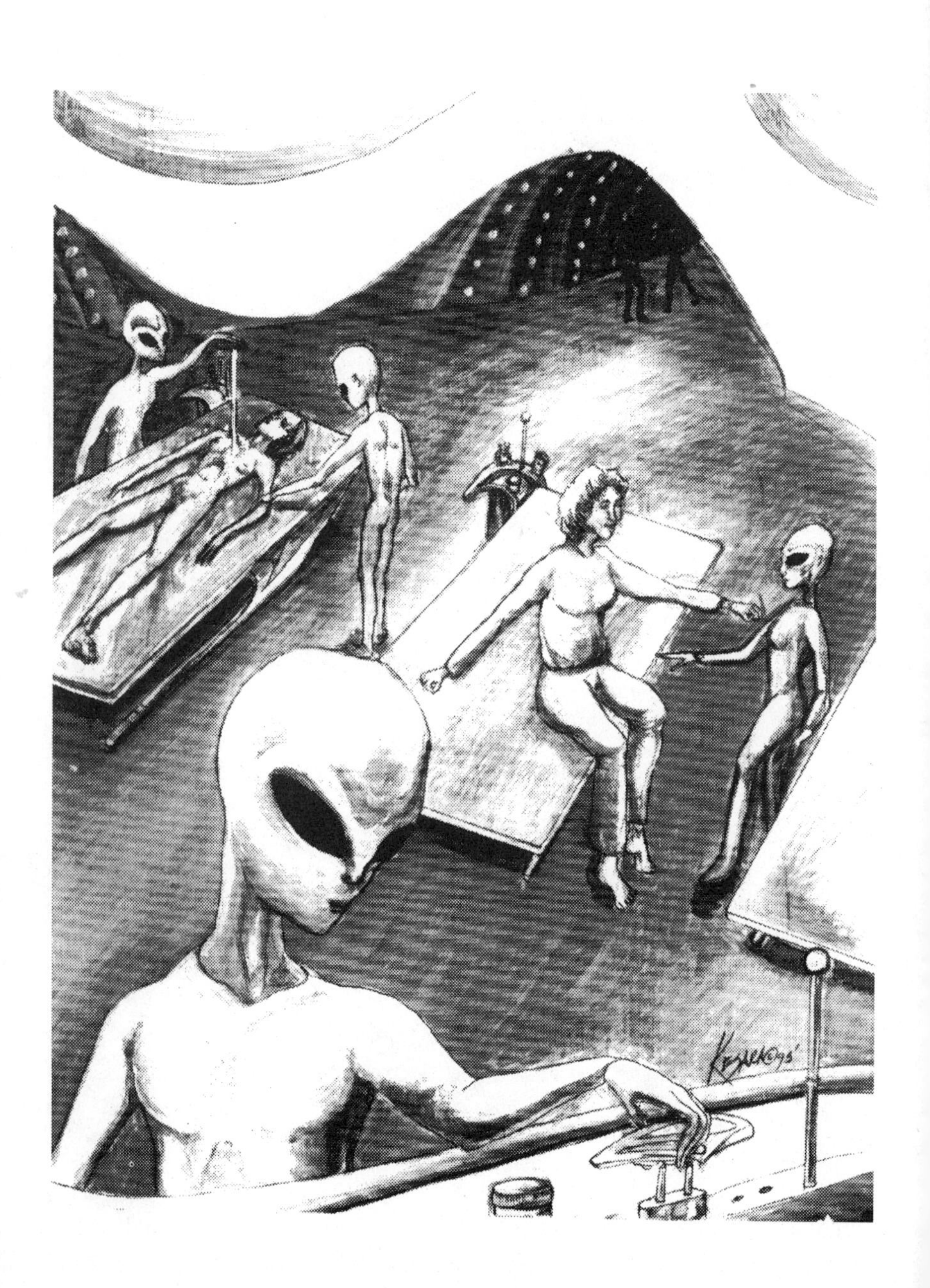

# Cures of Minor Illnesses and Ailments

Nearly every one of us has had the unpleasant experience of a bad cold. If you've never had a cold, perhaps you've had those nasty growths called warts. If not warts, the chances are very good that you have experienced a high fever or perhaps a backache, an earache or a minor infection of some sort.

At the state we are in right now, illnesses are simply a part of life. Minor illnesses are, by definition, not life-threatening. Unfortunately our doctors still remain helpless in the face of many of these minor illnesses. Many treatments exist, but few cures. Modern-day doctors, with their knives and potions, often are unable to help cure the sick person; sometimes all they can do is alleviate some of the symptoms.

Meanwhile, people are stricken with various illnesses everyday. Cold viruses spread through the population, causing millions of people untold suffering. The number of people dealing with one type of ailment or another is staggering.

As the medical field advances, more and more cures are becoming available. Still, the cures for thousands of minor illnesses continue to elude medical professionals. And so, until the field of medicine advances, people will continue to suffer from these minor ailments. Unless, of course, they get help from another source.

There are now many cases on record where people have received cures for all of the above mentioned illnesses, and many others. The following case is one of the few cases involving a person who was cured of a head-cold by a UFO.

CASE #035. On May 10, 1975, Chuck Doyle of Florence, Kentucky, went outside to check his horse. Suddenly, he heard a loud buzzing sound. Turning around, he saw a twenty-foot long metallic object shaped like a "manta ray" hovering over his neighbor's garden. The object had colored lights and was emitting a green beam of light towards the ground. Chuck was so stunned by the sight, he stayed perfectly still.

The beam of light began sweeping in wider circles, stopping for a moment on Chuck's pool. Then suddenly, the beam swung towards him. Chuck turned to run, but he was too slow and the beam hit his entire body. As Chuck says, "It was a straight shaft of light that didn't get wider at the bottom, like a laser...then the beam came at me. When it hit me, it was like being hit by a bucket of ice-water. I felt suddenly frozen. I couldn't move."

Chuck was actually frozen in the position of trying to run away. Although he was leaning forward, he was held in position by the beam. To his surprise, a flood of strange symbols and images filled his mind. He saw mathematical equations, images of a strange planet and a kaleidoscope of bizarre colors. Then the beam retracted and Chuck fell to the ground. The UFO disappeared with a purple flash and a loud bang. Chuck returned to his house in a state of shock.

It wasn't until he returned inside that Chuck noticed something strange. Before the encounter, he had been suffering from a miserable head cold. Immediately afterwards, all traces of the cold had disappeared. Chuck saw a doctor a week later, making no mention of his UFO encounter. The doctor pronounced him perfectly healthy.[1]

A cold, although not usually life-threatening, can be very hard to handle. Just look at the super-abundance of various cold remedies. Americans spend millions of dollars on throat lozenges, decongestants and pain relievers. Still, however, none of these medicines actually cure a cold. They only alleviate the symptoms, and even then, only partially. The only cure for a stuffy nose, a sore throat or a cough is time.

Unless of course, you are healed by alien influences, as this next case illustrates.

CASE #059. As mentioned before, Bert and Denise Twiggs of Hubbard, Oregon, are contactees who are friends with human-like aliens from Andromeda. On July 14, 1989, Bert Twiggs was bedridden with a severe cold. Denise begged him to go to the doctor, but Bert refused, saying that he would go if he didn't feel better the next morning.

---

1. Stringfield, 1977, pp. 63-69.

Bert went to bed, mentioning to Denise that his stomach had also begun to hurt. His cough was bad, and he had great difficulty breathing. As soon as they went to bed, they noticed strange lights in their room. These lights often appeared at the onset of an encounter. That night, they were to receive one of their most dramatic visitations.

At 4:00 A.M., Bert and Denise woke up with the memory of the alien house-call. Bert remembered the aliens arriving and being very concerned about his illness. They claimed that his condition was actually near fatal, and that they were, in fact, saving his life. They gave him an injection on his arm, and left. Only then did they allow the Twiggses to have full conscious volition. They both woke up, and Bert was not surprised to see a needle mark on his arm, exactly where he had been given the injection.

The next morning, his cold was noticeably improved and within forty-eight hours, it was completely gone.[2]

Some ailments are very persistent, and despite regular treatments, nothing seems to work. Many women know this from personal experience. There is a condition which plagues millions of women, which is known as candidiasis or yeast infection. Of course, this can be treated with antibiotics, however, the condition often reappears. The condition can be itchy and painful. Only recently have television advertisements presented medicine that is specifically designed for these infections.

There are at least two people who have stepped forward and said that their yeast infections were cured in a more unconventional way, that is, by extraterrestrials. As the next two cases show, the aliens are well aware of these types of infections.

CASE #097. Linda X., of California had been seeing a psychologist for three months before she discovered that she had been abducted by aliens. She went under hypnosis to recover the rest of her memories and discovered that she had been aboard UFOs many times, and so had her family. And not only that, they had cured her several times of many different illnesses.

During one of her encounters, Linda was cured of a persistent and nasty vaginal yeast infection. She was taken inside a UFO during the night and placed on a table. An instrument that looked like a "big cotton ball on a stick" was placed inside her. The aliens, who had

2. Twiggs, 1995, pp. 52-53.

small, pointed heads, large eyes and greenish skin, used the stick to apply a "jelly-like substance that looks like aquamarine blue, and it's clear and it's transparent." The substance felt very cold and the aliens told Linda that it would help to freeze the bacteria, so they could remove it.

Unfortunately, the infection eventually returned, which proves that even extraterrestrials are not perfect. However, they are still very talented. For as we shall see in future chapters, Linda was cured of cancer, and she and her family have all been trained by the aliens to become psychic healers.[3]

CASE #100. Linda X. also remembered seeing her sister, Sherry aboard the UFO, being cured of a yeast infection. Sherry was hypnotized to recall the blocked memories of her possible abductions and to help verify Linda's account.

With the aid of hypnosis, Sherry also remembered being cured of a yeast infection aboard a UFO. Her description is very similar to her sister's.

First, she was put on a table and a machine began to analyze her body. She was told by the aliens that they were looking to see if anything was wrong with her body. This helped to reduce her fear.

Her next memory was "being examined vaginally, and it seems like my feet were in some kind of stirrups or in something very similar to what you'd find in a doctor's office."

Sherry was surprised that at the time, her fear completely evaporated and she actually felt grateful that the aliens were helping her. Her next memory was of a cream being swabbed onto her infection. As Sherry says, "It was like a cream. It was put inside of me...real deep inside of me...it was a little cold and really messy. It felt...gushy. It wasn't really cold, but I could tell that it was like a cream."

Sherry was told that the treatment would take care of her yeast infection. The treatment seems to have been successful for the infection did not return.[4]

The advancement of medical science on Earth has been very rapid in the last hundred years. In the not too distant past, many illnesses were simply labeled as "consumption" or even worse, possession by

---

3. Fiore, 1989, pp. 89-113.
4. Fiore, 1989, pp. 114-131.

evil spirits. Barbaric methods of healing included blood-letting with leeches and prescribing all kinds of noxious tonics and potions.

When vaccinations became widely available, they were hailed as a great leap forward in medical science. Over the years, many medicines were discovered that helped eradicate disease. Penicillin and other antibiotics have literally saved countless lives.

Before these drugs, however, many people died from illnesses that are easily cured with modern medicine. One very common sympton of disease, fever, used to be a harbinger of death for many people if it remained too high. Today, fevers can be mitigated and controlled with a potent and relatively new drug called aspirin. Less than a hundred years ago, the popular method for controlling a fever was to encase the patient in ice.

Quite often people can suffer with fevers that arise from unknown diseases. Occasionally these fevers can be dangerously high, which can be fatal—unless, of course, there is extraterrestrial intervention, as this next case illustrates.

CASE #025. Jerry Wills of Kentucky had his first contact with extraterrestrials in 1965, when he was twelve years old. At that time, he would stay outside, night after night, watching star-like objects darting at right angles high in the sky over his house. A few months later, he had a close-up sighting of a typical flying saucer—a hovering metallic ship with blinking colored lights.

The next year, in the same location out in the woods, he was approached by a blond-haired blue-eyed stranger dressed in a beige jumpsuit and a silver belt. The man said his name was Zo, and that he was an extraterrestrial. And so began a very complex series of contacts in which Jerry was taught about science, history, philosophy, religion and other subjects. The aliens visited him three to six times a month for over five years.

The teachings seem to have been highly effective, for Jerry Wills has invented several items including a sighting and control helmet for pilots, a virtual reality device, and a "guardian crystal" which allegedly allows its user to detect auras.

Jerry Wills must have been considered valuable to the aliens. In addition to teaching him many things, they cured him when he was quite ill.

It was nighttime and Jerry was in his room suffering from a high fever. In the middle of the night, the aliens came and took him inside their ship. He was surprised to see the often-described grey-type aliens. They told him telepathically to "relax and let them do their work." He was given an injection in each arm.

The next morning, he woke up and found that his fever was gone. He recovered completely in less than a day. Although he has not had any other cures, Jerry still remains in contact with his alien friends.[5]

One myth about contactees is that there are no witnesses to support their accounts. Actually, in at least a quarter of the contactee cases, the contactee is asked to invite a group of people to witness the extraterrestrial craft. This bringing in of additional witnesses serves to strengthen the credibility of the contactee. Because of this, the contactee usually has more witnesses to support his/her testimony than does the standard abductee.

Many people are able to have UFO sightings this way. However, in the following case, a man was not only able to see a UFO, but was also cured of his illness simply because he was in the group of people hoping to see the UFOs of world-famous contactee Sixto Paz Wells.

CASE #055. In May of 1988, a professional male dancer was cured of his illness by a UFO. The event occurred at the Sebago Cabins Recreation Park in New York state.

The anonymous dancer joined about ten other people and contactee Sixto Paz Wells, at a predetermined location in the wilderness. After doing a lengthy cleansing meditation, the group watched several star-like objects move in darting patterns. The objects would also respond to the mental commands of the group members by moving according to telepathic directions. At one point, strange flashes of light illuminated the group. As one of the witnesses, Betty X., reports, "Another bolt of light hit our group, that for some strange reason made me dizzy and nauseous. Other people in our group reported the same sensation. The queasy feelings left me after a few minutes."

The group was told that they had received what the aliens called a "Xendra," which gives psychic and spiritual energy.

The anonymous dancer had come with his wife to the group. He was "very ill with his legs swollen and in pain." In the middle of the night, both of them were awakened by a "brilliant white light that illuminated the room and soon went away."

---

5. Randazzo, 1993, Chapter 2.

The next morning, when the dancer woke up, he was surprised to find his pajama bottoms on the floor. He didn't remember taking them off. It was then that he realized his legs were "pain-free and not swollen anymore."

Although the illness is not described in detail, it is evident that a healing took place. The other people in the group heard the man's story, and several others reported having contacts in the middle of the night.[6]

Another similar case involving the cure of an unidentified illness took place in New Jersey to UFO researcher, Ellen Crystall, best-known for her book, *Silent Invasion*.

CASE #008. Ellen Crystall is a well-known and highly respected UFO investigator who has taken UFO research into the field by trying to observe UFOs firsthand. She has been remarkably successful and has obtained numerous photographs and had multiple witness sightings. As a young two-year-old child, however, she had a more personal encounter.

As she says, "I had been sick for about nine months, but no one knew what I had. I recalled the doctors drawing outlines of my organs on my abdomen so my mother could point to the ones that were enlarging. I also remembered having to stay inside while my friends were outside playing...I asked my mother, 'What happened when I was sick? How old was I? What was going on?' She told me I was about one when I started getting sick, but no one knew why. She gave me a long account of how I got sicker and sicker...no one knew what I had except to say that my stomach was enlarged and very upset...my parents and the doctor feared I might die because I was so much sicker. I said to my mother, 'Then what?' She replied, 'One day in April you were perfectly fine.' It was very strange, she said. One day I was deathly ill and overnight I was completely better. She never figured it out."

Years later, Crystall was told by a psychic that she was healed by aliens who had hovered outside her home and effected a cure. The psychic provided other information that was verifiably accurate, such as the color of her house and the date of illness.[7]

---

6. Randazzo, 1993, Chapter 3.
7. Crystall, 1991, pp. 5-6.

There are many ways to interpret a UFO experience. In the following case, the witness was visited by an entity which healed her of a chronic back condition. Mark Chorvinsky, editor of *Strange Magazine,* wrote about the case and remarked upon the many parallels to the classical accounts of the "Grim Reaper" in which witnesses see a "hooded, robed, faceless entity." A close examination of the case, however, reveals many details that are in accord with UFO accounts.

CASE #030. In March, 1973, Olga Adler of the United States was lying in bed suffering from "chronic back pain, making any movement, especially bending, painful and difficult."

Her eyes were closed when the room suddenly became bright. Adler was just as suddenly paralyzed. She then perceived a "figure in a long, light brown robe, resembling a monk's habit, with the hood pulled up and rendering the face in deep shadows."

The figure entered through a closed door and floated across the room. At this point, the entity used an instrument to cure Adler of her chronic back pain. As she says, "In his arms he held what looked like a heavy metal cylinder behind my back as I lay there on my side, and pressed it down. I could feel the bed depress with the extra weight and could see the figure leaning on it directly behind me, the face always in deep shadow. With contact of the cylinder at my back, I could feel a comfortable warmth penetrating my body and felt surrounded by a tingling sensation as electricity. The figure bent over me and held the cylinder like that for at least five minutes, as I enjoyed and relaxed in the comforting warmth."

At this point, the figure departed through the closed door and Adler fell asleep. Upon awakening, Adler was astounded to discover that her backache was gone. As she says, "It was completely gone, and I could bend and twist and move with absolutely no pain…I believe I had been healed by some kind of spirit guide or angel."

One year later, Adler had another encounter in which the figure entered her bedroom and spoke telepathically, "Come with me—it's time to go!" Adler has vague memories of being floated outside through a closed window. From that point, she has missing time.[8]

In the following case, another lady was cured of chronic back pain during what was initially a frightening encounter with a UFO.

---

8. Chorvinsky, Oct., 1993, pp. 22-24.

CASE #070. Late one evening in 1991, a twenty-one-year-old anonymous female was driving with her sister through the outskirts of Claremont, California. For years she had suffered from intense chronic back pain. As they were driving, they both spotted a disk-like craft hovering above them, casting down a bright beam of light. Barbara Lamb reports what happened next: "The car stopped. So they got out of the car and started running across the field. And they were followed by a beam of blue light. And the beam suddenly struck her in her lower back and she felt it radiating right through her, and felt a very, very powerful energy."

The main witness was running in terror from what she thought was going to harm her. Ironically, it had the opposite effect. As Barbara Lamb says, "As a result of that, she felt that she had been healed of whatever the difficulty was and was actually quite grateful to them."[9]

A multitude of different illnesses plague humanity. In our efforts to combat these illnesses, we take medicine. Unfortunately, many medications cause unwanted side-effects. In the following case, a lady was suffering from dizziness brought on by the medicine her doctor had prescribed. It may be the only case on record in which aliens cured someone of side-effects of terrestrial medication.

CASE #067. An anonymous nursing home worker from Deming, New Mexico reports that in 1989, she was suffering badly from dizziness as a side-effect of medication. The dizziness was so bad that for three months, she could only walk by holding onto furniture.

During that time, she had a dream-like experience in which she saw four men, each four feet tall with coveralls and hoods over their faces. They were holding a "mysterious box." When the witness felt herself slammed back onto the bed, the entities were gone. The witness was sweating profusely. She went to the bathroom where she discovered that she was cured of her dizziness. As she says, "I think those little men took away my dizziness. And I am grateful for that. I think that they will come whenever I need them."

The witness reports that right after going to the bathroom, a heavy object struck her roof and ran across it. She believes it is related to her UFO experience. She has also had several UFO sightings in the past.[10]

---

9. Personal files.

Some UFO investigators have noticed that there is a high incidence of sinusitis among abductees. Some investigators speculate that this is a result of nasal implants. This may or may not be true. However, there is at least one case on record in which an abductee was cured of chronic sinusitis as a result of her abduction.

CASE #088. Ann X. from Chicago suffered from chronic sinusitis all her life, until she had an onboard UFO experience. Ann's experience was typical. She was placed in a room where she underwent an examination at the hands of alien "doctors." She saw full-color living x-rays projected on a screen on the wall of the room. She describes the experience in familiar terms. As she says, "The 'doctor,' who was not visible, spoke in a high-pitched, lilting, pleasant voice as I lay in a comfortable, high-tech 'dentist's chair.'"

At this point, the aliens performed a cure. As Ann says, "One of the alien figures standing near me placed a wand near my nose and cured my sinus condition.... Curing my sinus problem, they said, was their gift to me of good faith."[11]

There are many other cases on record which lack the wealth of detail of the above reports. Despite their brevity, the reports are still quite impressive.

CASE #101. Abductee Linda X. saw many other human beings being operated upon by the aliens. One man was strapped to a table, with a narrow metal instrument hovering over his midsection. When Linda asked what was wrong with the man, they told her that he had a stomach ulcer. They were curing the ulcer by generating energy and directing it to the proper location.[12]

CASE #102. Abductee Linda X. saw a young girl being examined on a table by the aliens. They performed the standard physical and discovered that the girl suffered from intestinal worms. To cure the girl, they injected her abdomen. As Linda says, "It looks like a big syringe in the stomach. They put salve in and carried her off the table."[13]

---

10. Leupold, Jan. 27, 1994, p. 4.
11. Steiger and Steiger, 1994, p. 43.
12. Fiore, 1989, p. 105.
13. Fiore, 1989, p. 105.

CASE #104. Abductee, Linda X. saw a large, dark-haired man being examined by the aliens. They told her that the man suffered from a bursitis-like condition in which his left shoulder dislocates and becomes swollen. They tested the mobility of the arm by raising it. Finally, they cured his condition using an instrument with a "pulsing light."[14]

CASE #103. Abductee, Linda X. saw a young, teen-age boy lying on the aliens' examination table. She was told that the boy suffered from chronic hip problems on his right side, and that the bones were deteriorating. He walked with a pronounced limp. Linda watched as the aliens placed a large L-shaped instrument over the boy. It emitted green, yellow, blue, pink and white light, which Linda was told would "promote new cell growth." After the operation, he got off the table, still limping but much improved.[15]

CASE #096. A man from Tblisi, Georgia, in Russia was operated on by friendly extraterrestrials who cured him of his chronic back pain.[16]

As we have seen, minor illnesses and ailments of many types have been cured. The cures have been made with beams of light as well as surgical operations and injections. Several of the cures were verified by doctors or other credible witnesses. Despite the controversial nature of some of the accounts, the evidence that these cures have taken place is overwhelming.

The question, again, is why? Would extraterrestrials really travel millions of miles to cure a person's cold? Would they really abduct somebody for curing something as minor as a yeast infection? Are they really concerned about a little girl's case of intestinal worms?

Evidently, yes. Despite the fact that these accounts are hard to believe, it is only logical that such cures are possible. After all, the aliens have exhibited an extraordinary knowledge of the human body. Curing human ailments appears to be a simple and routine event, often taking only minutes to perform. People's bodies are open and closed as if they were equipped with zippers for easy organ access.

Up to this point, we have examined cases of injuries and minor illnesses. Now we move to a more specialized area of medicine, that

---

14. Fiore, 1989, p. 105.
15. Fiore, 1989, pp. 105-106.
16. Morrow, Jan./Feb. 1993, p. 17.

of opthamology. As we shall see, the aliens are very knowledgeable concerning the health of the most sensitive of human organs: the eyes.

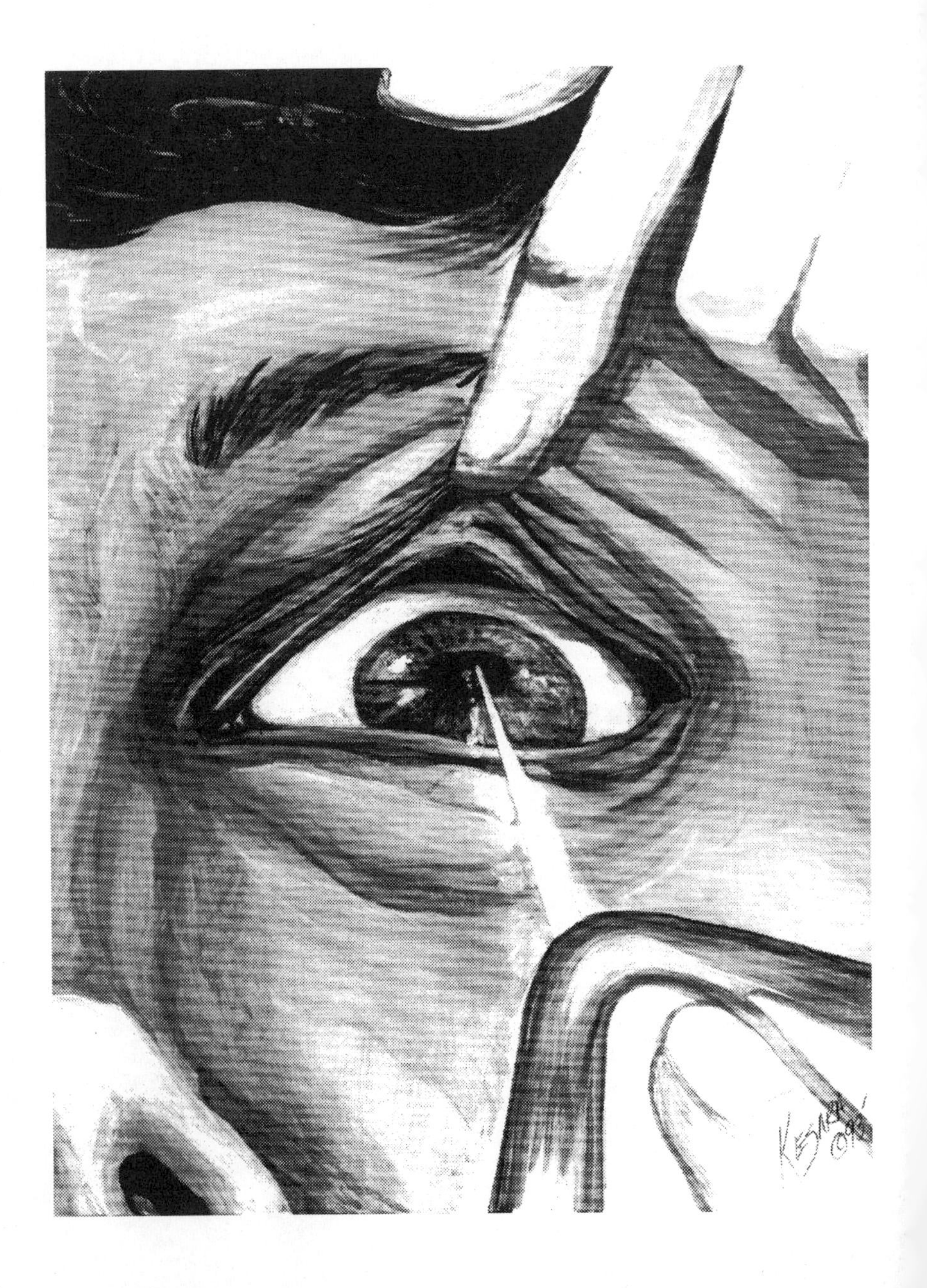

# Eye Doctors From Outer Space

The eyes are one of the most important organs of the human body. Although not essential for life, over ninety percent of all sensory information comes through the eyes. The eyes are often described as the window to the soul, and rightly so. Even the aliens are known for their large, dark eyes which seem to see right through the frightened abductees.

The list of known eye disorders is long. Some of the more common conditions include near and farsightedness, astigmatism, eyestrain, cataracts, sties, glaucoma, strabismus (cross-eyes), detached retina and countless other lesser-known disorders.

Because of these conditions, eye doctors advise their patients to have regular six month check-ups. The field of eye medicine has had many recent strides. Many vision problems can be corrected through relaxation drills or simple corrective lenses. Another method is called radial-carototemy; by surgically changing the shape of the eyeball, doctors are able to restore clear vision. Cataract surgery can now be performed on an out-patient basis. Crossed eyes can be corrected with vision exercises or, in severe cases, corrective surgery. Even detached retinas can be cured with laser surgery, or in severe cases, by actually removing the eyeball and sewing the retina back onto the eyeball.

Despite these recent advances, or perhaps because of them, eyecare is a million-dollar-a-year industry. Over fifty percent of Americans wear corrective lenses. Thousands more are diagnosed and treated for various disorders each year.

Some fortunate people, however, have received seemingly miraculous cures from UFOs.

The following case which has been recounted in many UFO books and articles is without a doubt the best-known of all eye-cures performed by extraterrestrials.

CASE #022. On December 9, 1968, a Peruvian Customs official saw a UFO outside his home. It was hovering quite high up in the sky and appeared to be disk-shaped. As he watched the object from his terrace, it emitted a long, thin beam of light. Before he could react, the Customs official was struck on the face by the "violet rays that it emitted." While in the beam, the official was unable to move.

Before the encounter, the anonymous official suffered from severe myopia that forced him to wear thick glasses. However, after the encounter, he was able to see perfectly without glasses. Even more amazing, the man had suffered from rheumatism, but after being struck by the beam of light, his rheumatism was gone. As we have seen, he is not the only person to receive a multiple cure from a single incident.[1]

The types of procedures performed on some abductees proves that the aliens are quite knowledgeable about our eyes. One perfect example is the chilling testimony of famed abductee, Betty Andreasson. During one of her many abductions, the aliens physically removed her eyeball from its socket and probed behind it with a large needle. The experience, although painful, seemed to cause no permanent damage. Betty's vision remains as it was before the incident. It seems that the operation was performed to insert an implant rather than to perform a healing. Betty has 20/20 vision.

Leah Haley, abductee and author of *Lost Was The Key*, also reports having an experience identical to Betty Andreasson's. During one abduction, Haley reports, "They're doing something with my right eye. It seems like they're reconstructing the whole thing. It doesn't hurt, but I feel pressure...when they're messing with my eye, it feels as if they're inserting an object into my nose. But they don't do it from the nostril; they put it in the eye area...it doesn't bleed. But after they take the eye out and put the object in and put the eye back in, my nose hurts."[2]

Late abductee, author and UFO researcher Karla Turner, Ph.D. was familiar with these types of cases. As she said, "Abductees' eyes

---

1. Blum, 1974, p. 147.
2. Haley, 1993, p. 150.

are painfully removed from the sockets, allowing the aliens to implant devices into the area before the eye-balls are replaced, for instance."[3]

Most alien eye operations are not so unpleasant. Take the following case of an eye cure performed on the son of two contactees.

CASE #064. In December 1989, Christopher Twiggs, the son of contactees, Bert and Denise Twiggs of Hubbard, Oregon, was taken aboard a UFO with his family. The reason for the visit? It seems that Christopher had gotten a sty in his eye. His mother knew about the sty and had plans to take him to the doctor the following day. However, the alien doctors beat her to it.

Unfortunately, there are no details on how the cure was actually performed. Christopher was simply sent to the examination room for treatment. As Denise Twiggs says, "After a trip to the medical room aboard ship, his eye was nearly healed the next day."[4]

One very interesting case of a eye healing comes from UFO investigator John Carpenter and involved a man who was cured of profound color blindness.

CASE #089. This case involves a 21-year-old man from the midwestern United States by the name of Eddie X. Eddie was in the middle of a series of hypnotic regressions to explore his many abductions when he experienced yet another encounter. As John Carpenter says, "During a physical examination by the beings, Eddie's right eye was removed. He felt it pulled out and replaced. Afterwards there was a redness, a physical soreness. He did not feel they were implanting anything behind his eye, only that they were 'fixing' him. The side benefit of this operation appears to be enhanced colors. The procedure was clearly for their purposes, not for Eddie's, but the partial accidental cure of his lifelong color blindness was a side effect.

Carpenter further substantiated the healing by obtaining a statement from Eddie's eye doctor which says that Eddie suffered from "profound color blindness," though after his experience, his vision "had improved up to green color blindness."[5]

Poor vision is often a nuisance. Blindness, however, can be devastating. Although rare, blindness afflicts a significant portion of the

3. Turner, 1992, p. 12.
4. Twiggs, 1995, p. 130.
5. Bryan, 1995, p. 26.

population. It is very rare that cases of blindness can be reversed. Most blind people must learn to cope with their disability. Some are lucky enough to have a seeing eye dog. They all, however, go through life in total darkness.

As we have seen, contactees are often healed of their illnesses and injuries, presumably so that they can fulfill the missions given to them by the extraterrestrials. The following case is a perfect example. It is also one of the earliest recorded cases of a UFO healing.

CASE #004. In April of 1945, on the island of Okinawa in the Pacific Ocean, contactee and professional painter Howard Menger was tossed right into the middle of World War II. Menger was on one of the aircraft that saturated the Japanese-held island with bombs. After dropping hundreds of bombs and peppering the shore with machine-gun fire, the Americans were able to take control of the island.

Still, the Japanese continued to attack. Some had managed to survive the bombing by hiding in caves. These men would continue to attack in the style of guerrilla warfare. Also, a nearby Japanese-held island shelled Okinawa constantly. Although it caused little real damage, it was one of these shellings which proved disastrous for Menger.

One day, as he patrolled the airstrip, another shell from the nearby island fell short of its target and landed on the airstrip. Menger heard it coming, so he fell to the ground and stayed low. After the explosion, he was relieved to have escaped injury. Then he felt something hit his eye. As he says, "As I got to my feet I felt something stinging in my right eye. I put my hand to my eye and managed to pick something out with my fingers. It was a piece of shrapnel."

Menger rushed to the medical tent where he received treatment. Unfortunately, the eye became badly infected. Before long, Menger was totally blind in his right eye.

Menger was, of course, unable to perform his duties, and remained in the medical tent under the care of the overworked doctors and nurses. Unfortunately, Menger's eye infection was very severe, and it quickly spread to his other eye. Before long, Menger was totally blind in both eyes.

He was very upset and kept the news from his family back home. He also remembered what his extraterrestrial friends had told him. Ever since he was a little boy, Menger had been contacted repeatedly by human-like aliens who claimed to be from various planets in our solar system. Through the years, Menger had many typical contactee adventures. He was taken board the UFO and given a tour of the moon. He was given instructions about how to build an energy-free motor. He was told many predictions which later came true. He was able to take his friends to see the UFO and was provided with further demonstrations to prove the existence of his aliens friends. He was even allowed to photograph the UFOs, the aliens, and his trip to the moon. Many of these photographs appear in the book he wrote about his experiences. But most important here, Menger received a healing from the aliens.

As he says, "Something happened in the hospital tent that I have often wondered about. Perhaps I can never be certain." Menger states, that during his stay in the hospital tent, he was visited by an unidentified and very kind lady. When Menger asked if she was a nurse, she didn't reply and told him that he was the one she had come to see. Menger was struck by the fact that she seemed to know a good deal about him.

Although no specific medical treatment is mentioned, Menger had a sneaking suspicion that she was one of his extraterrestrial friends, and that she was sent to help him. As Menger says, "She assured me my sight would be restored."

Very gradually, Menger's sight returned and he was able to behold the lady who had helped him. He saw an "attractive woman with wavy brown hair, dark eyes, and fine white teeth." She was dressed in the normal Army nurse's uniform.

Menger had further reason to believe that the lady was actually an alien. As he says, "Although I suspected she was one of the space people, she never made herself known directly. Near the time of my release she said that I would soon meet a very interesting person. I assumed it would be another contact."

The prediction made by the mysterious lady turned out to be correct. About two weeks later, he had a strong impulse to drive to a remote area on the island. He did so and there he met a human-looking extraterrestrial. The ET told him that there would soon be a terrible bomb. And although the ET didn't mention Menger's recent healing, he did explain why they did it. As the alien told Menger, "We have been spending a lot of time conditioning you and preparing you for your work to come. We are contacting people all over the world."[6]

---

6. Menger, 1959, pp. 39-47.

Two of the most amazing extraterrestrial eye cures happened to two ladies who were abducted together as children, separated, and then brought together through an amazing series of circumstances. The two ladies both feel that they were brought together by the aliens. The case was investigated by Budd Hopkins who says, "They were extremely impressive, articulate, obviously filled with a lot of emotion about their experiences, a lot of anger, a lot of fear. I was just extremely impressed with them."

CASE #073. Beth Collings of Virginia had experienced alien abductions throughout her childhood. As an adult, she continued to be bothered by memories of abductions. In 1987, she met Anna Jamerson and began working at her horse farm. The two ladies didn't know it, but as children, they had been abducted together. When they met, they both felt that they knew each other. It wasn't until much later that they discovered their shared childhood abduction.

Meanwhile, Beth continued to have encounters. In December of 1992, she experienced a terrifying missing-time abduction. All she remembered was seeing a "low flying aircraft." Then her car engine shut down and a bright light was emitted over the car. Beth remembered being "frozen to the spot." Her next memory was of racing home, realizing that she had lost an hour or more of time.

Almost immediately after the experience, however, she noticed something strange. She had bad eye sight before the experience. Afterwards, her vision was perfect. As she says, "The first thing that I noticed from that first experience was that my contact lenses appeared to be missing from my eyes, and I certainly had them in when I was driving." Beth then realized that her vision had improved dramatically. Beth eventually sought a UFO investigator and underwent hypnotic regression to recover her memories. Several of her memories of abductions are extremely unpleasant, involving painful examinations and procedures. After one abduction, the aliens placed her outside her home on the ice, instead of back in her bed from where they had taken her, causing all the skin on the bottoms of her feet to be badly burned.[7]

CASE #074. Anna Jamerson was surprised by Beth's encounters and was even more surprised that Beth's vision had improved as a result of the latest experience. At the time, she had no idea that she her-

7. Collings and Jamerson, 1996.

self had been abducted as a child and was soon about to have another abduction. In fact, her only clue that she had had a UFO experience was a miraculous cure almost identical to Beth's. As Diane says, "My eyes had improved dramatically and I could think of no other rational explanation why my eyesight was also improving unless I too was being abducted." Anna underwent hypnotic regression and discovered that she had, in fact, been abducted on several occasions.[8]

These accounts complete the record of the UFO eye cures. There is another case, however, of a more bizarre nature. Although it is not strictly a healing, this case involves a lady who developed a strange eye disease after having a UFO encounter. Shortly after her encounter, she became legally and irreversibly blind. Later, seemingly because of her encounter, she began to develop extrasensory sight. It remains one of the strangest healing cases on record.

CASE #009. In July of 1952, eleven-year-old Marianne Cascio Shenefield of Agawam, Massachusetts, went outside to play. She was alone when a metallic object flew over her head. A few minutes later, she noticed a figure standing next to her. The figure had gray skin and large almond-shaped eyes. He was wearing a metallic-looking jumpsuit and a black belt with a silver box and colored buttons on it. Before she knew it, Marianne was abducted.

She remembered standing in front of a screen that showed her internal organs pulsing with life. Marianne promptly vomited upon the alien standing next to her. They quickly put a tube to her mouth which seemed to give her oxygen. After examining her, they gave her a smooth, black stone with writing on it. Marianne was then allowed to exit the craft.

Almost immediately after the encounter, Marianne's vision began to deteriorate. An eye doctor diagnosed her with the genetic disorder known as Stargarrdt's disease, which is a degeneration of the macula. The end result is the destruction of the central vision and portions of the peripheral.

In 1972, Marianne heard strange beeping sounds outside her home. She went outside, and to her surprise, experienced another UFO abduction. Two years later, she began hearing the beeping sounds in her ears. Then on March 15, 1974, Marianne's sight suddenly and inexplicably returned for a brief interval of ten minutes. Mari-

---

8. Collings and Jamerson, 1996.

anne called the doctor who told her she must be hallucinating as her condition is irreversible.

Then Marianne began having vivid, colorful visions. She saw formulas, equations, genetic-codes, colorful geometric shapes, numbers and letters. Her house also became plagued by poltergeist-like phenomena. When the doctors couldn't help, Marianne went to a priest, who was also unable to help. Finally, she hooked up with a paranormal investigator and began to come to terms with her condition.

Today, Marianne's psychic sight is very well developed, and in many ways, she "sees" better than the average person. Marianne is also convinced that her vision is the way it is today because of her UFO encounters. She is not pleased to be blind and in no way does she consider her condition a healing. However, she does admit that her psychic sight is a result of her encounters. She doesn't know if the aliens caused her blindness or whether her loss of sight was coincidental. It does seem possible, however, that the aliens were aware of her upcoming blindness and may have simply repaired her sight in the best way they could.[9]

As we have seen, aliens have cured numerous types of eye disorders. They have abducted people, shone down beams of light, or simply visited people in the hospital—all in order to cure people's vision problems.

So far we have studied healings of injuries, minor illnesses and eye problems. The human body, however, is an enormously complex, biological machine involving numerous organ systems that are only partially understood.

The aliens, however, seem to understand the human body better than we do ourselves. In the next chapter, we shall see how the aliens have performed cures on one particular organ system of the human body—in this case, the integumentary system.

---

9. Sable, Feb./Mar. 1991, pp. 14-18, 66.

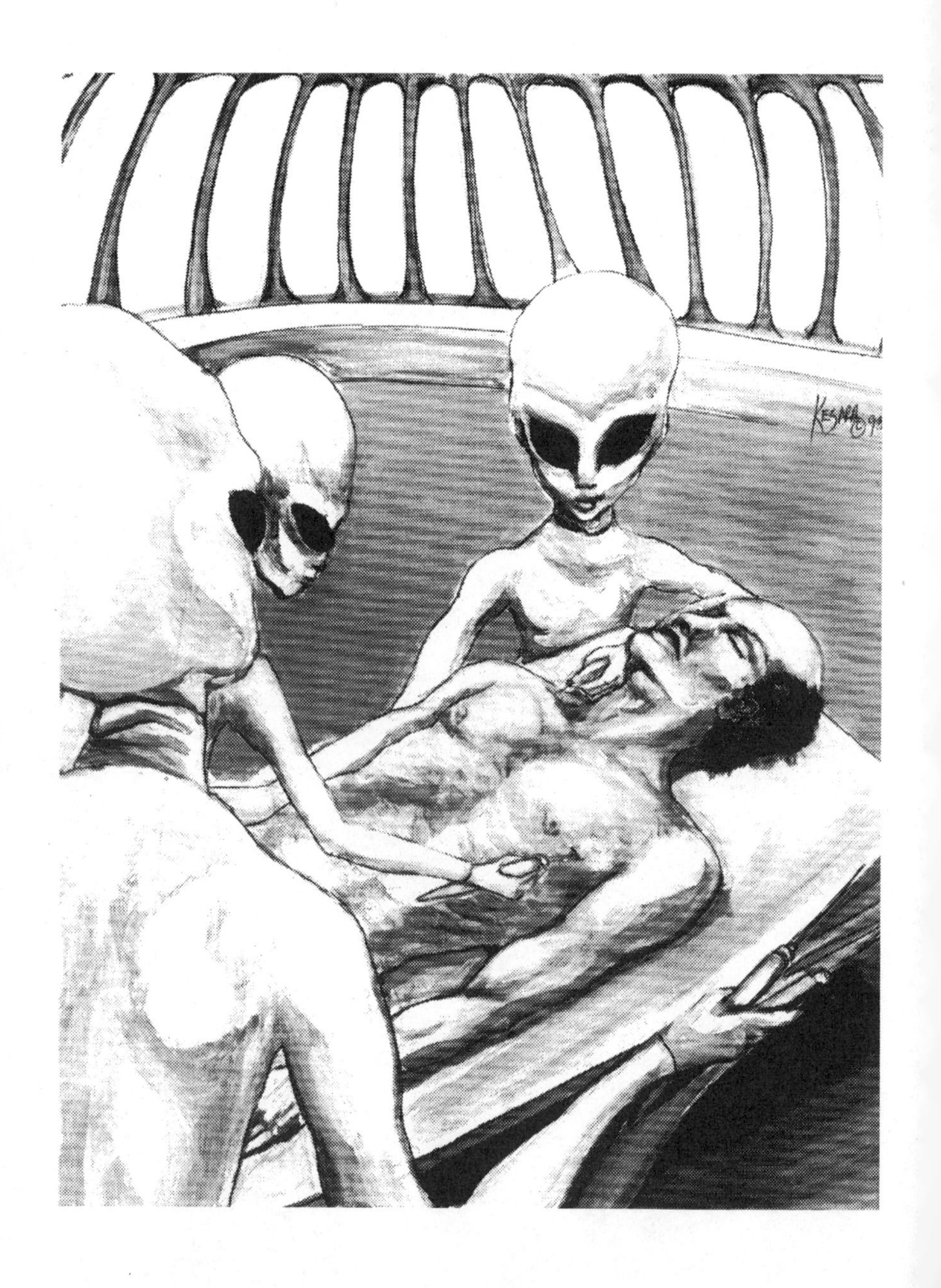

# Cures of the Integumentary System

The human body is composed of various organ systems that cooperate to keep the body alive. The blood system, the nervous system, the respiratory system, the immune system and several others all perform vital functions. Each organ system is essential for life. A breakdown, can lead quickly to death.

The largest organ system in the human body is known as the integumentary system. This includes hair, skin, teeth and nails. It performs the function of covering the body and protecting it from infections. Like all organ systems, it is vulnerable to a number of illnesses.

The integumentary system, however, has one distinction that none of the other organ systems have; it is responsible for peoples' appearances. For this reason, the integumentary system is of great concern to a large number of people. This organ system receives much more attention than most other organ systems. This is evidenced by the fact that Americans spend more money on skin, hair, teeth and nails than on the United States Space Program. In other words, we spend more money on the integumentary system than we do on the solar system!

Despite this fact, the integumentary system remains largely a mystery. People are afflicted with rashes, acne, hair-loss, weak nails and countless other conditions for which there seems to be no cure. As we age, our skin dries and wrinkles, and doctors remain helpless to halt the process.

Aliens, however, have performed some of their most remarkable cures in this area. In fact, they seem to work quite easily with this particular organ system. This next case is one of the best-known UFO healings and has been widely circulated in the UFO literature. It also represents one of the strangest UFO healings on record.

CASE #029. On December 30, 1972, nightwatchman Ventura Maceiras of Argentina saw a bright light hovering above his home. Inside the light, he could discern a metallic craft with portholes. Through the portholes, he saw humanoid figures. At this point, Maceiras was struck by a "brilliant flash of light" which shot out from underneath the craft.

For weeks following the encounter, Maceiras suffered from symptoms typical of radiation poisoning, including, hair-loss, headaches, nausea, diarrhea, eye irritation and "swollen red pustules" on his neck.

Doctors were baffled, but when Ventura told them of his UFO encounter, they had no choice but to believe him. His symptoms simply could not be faked. Besides, there were other details that supported Ventura's story. The tops of the Eucalyptus trees were scorched where the object had hovered. Also, Ventura's cat disappeared in full view when the beam came out of the craft. Forty-eight days later, the cat reappeared with severe burns on its back. Finally, an abnormally large number of dead catfish were found in a small stream at the location of the sighting.

In fact, there was so much evidence, Ventura's case attracted a lot of official attention. In less than a month, Ventura was interviewed over sixty times by various officials including doctors, police, government officials and others.

Ventura took weeks to recover from his injuries caused by the UFO. Then, less than two months after the encounter came the final but most unbelievable symptom of all. Verified by doctors, engineers, police officers and professional UFO investigators, Ventura Maceiras, although seventy-three years old, began to grow his third set of teeth!

UFO investigator Pedro Romaniuk was assigned to the case because of his high credentials as a former commander with an international airline and technical investigator for the Argentine Air Force Aviation Accidents Investigation Board.

Romaniuk performed a follow-up visit to Ventura, to check on his new growth of teeth. As Romaniuk says, "Since approximately February 10, Maceiras has observed that new teeth have been appearing in his upper left gum. At the time of my visit, I was able to confirm that two front teeth and two cheek teeth were coming through and were approximately 2 mm to 3 mm long."[1]

---

1. Blum, 1974, pp. 143-145, and Stringfield, 1977, pp. 72-74.

There is a bacterial condition that afflicts many people for which there are many treatments but no known cure. The condition is known as warts. Currently, there are several popular treatments. One is to apply a lotion that, after repeated applications, dissolves the wart. Another is to use a beam of light known as a laser and burn it off. Another new method is to actually freeze the wart off the body. There is also some evidence that warts can be reduced with mental exercises such as visualization. It is also true, however, that warts can fall off spontaneously, leaving no traces of their presence.

Still, millions of people suffer from warts. There is at least one incident, however, of warts being cured by aliens.

CASE #061. On September 4, 1989, farmer Jan DeGroot of Amsterdam, Netherlands, returned to his home to find one of his greenhouses glowing with strange light. He parked his car and approached the greenhouse and found a "large flat-topped disk" glowing with green light parked behind the greenhouse.

Suddenly, Jan was approached by a human-looking man with dark hair and straight features, dressed in a dark-colored jumpsuit. The stranger told him that his tulips were over-watered."

Jan's next memory is waking up the next morning, and feeling that his encounter was probably more extensive than he remembers. He instantly went to the greenhouse where he found a leak in his irrigation system, proving that the tulips were being over-watered.

He also received further proof of his encounter. For many years, Jan had suffered from a very large wart on his neck. The ugly wart had been there for so long that Jan was accustomed to it. It wasn't until the day after the encounter that Jan made the remarkable discovery: all traces of his wart were gone. He attributes the healing to his encounter with the UFO.[2]

Warts are only one annoying skin condition that seems to resist healing. Another condition which is absolutely normal is moles. Moles are simply a small defined area of pigment. Often hairs grow out of moles. Many people find them annoying and have them removed with a very simple laser procedure. Also, some moles are prone to become malignant and doctors sometimes recommend that they be removed.

---

2. Tessman, Winter 1993, pp. 62-63.

The following case may be the only recorded account of a mole being removed by extraterrestrials.

CASE #049. In August of 1982, farmers, Carl and Dagmar R. of northeast Iowa were in their home when they saw a strange light in the sky. In the days that followed, the couple found several circles of burned grass about thirty feet in diameter. One evening, Dagmar saw a figure with a strangely-shaped head and large dark eyes, out in the field.

In October, Carl was outside when he saw the same strange light hovering overhead. At three A.M. the next morning, Carl was awoken by the sounds of his cattle bellowing. Looking out the window, he saw a landed disk-shaped object glowing with green light. He woke Dagmar in time to see three entities enter through the bedroom door.

The two were quickly and quietly taken inside the UFO. Both of them underwent a physical examination and had reproductive material removed. Dagmar also remembers her hair and fingernails being clipped. She was rubbed with a strange lotion. Shortly later, she was alarmed to detect the odor of burning flesh. To her surprise, the aliens were removing a mole under her left arm. A short time later, the young couple was returned to their home. The next morning, they remembered only parts of their experience, though their memory returned naturally after a few days.[3]

One condition that affects all people is known as aging. One by one, our organ systems slow down and begin to shut off. Of course, this is true for the integumentary system. Skin cells don't regenerate as quickly and wrinkles form. Cuts takes longer to heal. Hair becomes gray, and for many people, pattern-baldness sets in. The teeth become yellowed and warn. Nail growth slows down.

Of course, we do all we can to rejuvenate our integumentary system, some even going as far as plastic surgery. But human doctors are not alone in their experimentation. There is one very famous case of the rejuvenation of a man's entire integumentary system, This case is very well-known and is the most amazing integumentary cure on record.

CASE #054. On March 20, 1988, professor of Indian Studies at the University of North Dakota, John Salter, Jr., and his son were driving

3. Steiger, 1988, pp. 97-103.

near Richland Center, Wisconsin, late at night when they both experienced a typical missing-time UFO abduction.

Under hypnosis, the standard scenario of an alien physical examination was unveiled. John and his son were able to corroborate each other's testimony. John remembered being injected several times by long needles as he lay on the examination table.

At first, there was no effect from the abduction other than the hidden trauma. However, almost immediately after the encounter, changes began to take place.

Salter used to smoke a pound of pipe tobacco every week. After the encounter, he slowly lost his desire to smoke and by May, he had stopped smoking completely.

Other changes were also immediately apparent. John Salter, Jr., reports that he had physiological improvements in over eighteen areas. These began showing up immediately after the encounter, and continued to manifest in the days and weeks that followed.

Some of the more significant changes are "improved skin tone, circulation, eyesight, faster blood-clotting after cuts or scratches, and faster and thicker growth of toenails and hair."

Salter also reports more improvements. As he says, "...many facial wrinkles have disappeared and my skin is much healthier; some scars have disappeared and others are fading; my face has narrowed and my neck area has slimmed (as has my entire body to some extent) and none of this has involved 'sagging.' My circulation has improved. Cuts clot immediately and heal very quickly; my energy level is way up; and I've had no colds or flu or any other illnesses since March 1988."

Salter and those who know him have no doubt that these improvements are taking place. The improvements are obvious. As he says, "After all these years, I have a five o'clock shadow."

John Salter III also report that his cuts heal at a very rapid rate.

Today, Salter Jr. has been interviewed by UFO researchers and has appeared in several videos. He also lectures at UFO conventions across the country about the beneficial effects of his abduction. [4]

This concludes the record of UFO healings of the integumentary system. Of course, logic would dictate that there are many other cures that go unreported. I know of one case involving the cure of warts and another involving the cure of acne by a beam of light from a UFO.

---

4. Schmidt, Dec. 1989, p. 6. and Salter, Vol. 1, No. 1, pp. 1-3.

Unfortunately, I was unable to locate the sources for these two accounts, but they do indicate the fact that more healings have occurred. We have also seen in other accounts how the aliens have the ability to stitch together a cut with a beam of light. And finally, there is an account of a healing of skin cancer that will be covered in a later chapter.

As we can see, the cures, as usual, border upon miraculous. The human doctor who could cure a person's integumentary system the way the aliens have would be an instant millionaire. Unfortunately, we are not quite as advanced. Therefore we continue to pour money, time and research into the integumentary system. Hopefully, we will some day approach the level where the aliens are now.

# Liver and Kidney Cures

The liver and the kidneys serve very similar functions. The liver, like the heart, used to be thought of as the seat of emotion. Today, doctors have adjusted their theories. The liver is the largest glandular organ in vertebrate animals. The liver secretes bile, and helps to metabolize carbohydrates, fats and proteins. It also serves to filter the blood, and is notorious for collecting toxins from the blood system. As we all know, alcoholics often drink so much that their livers become badly damaged. Their livers are simply unable to keep the blood clean. A liver is, of course, a vital organ—meaning, the human body would die without one.

The kidneys, of which there are two, are also vital organs. They serve to secrete urine from the body. Kidneys are glandular organs composed of minute tubes lined with cells which separate water and waste products from the blood.

Both livers and kidneys are susceptible to a number of maladies too numerous to list. People suffer from everything from liver spots to kidney stones to total organ failure.

Cures for the liver and kidney have improved over the years. For those serious cases, organ transplants are available. Kidney stones are often treated with a new device that uses ultrasonics to break down the stones, making them easier to pass with the urine. For serious cases, there is dialysis and other life-support machines.

These two organs have been the focus for an extraordinary number of UFO healings. The following case is only one of many cases we shall examine. Like many UFO healings, it is authenticated by the examining physicians.

CASE #020. In 1967, contactee Ludwig F. Pallman was hospitalized in the Maison Francais Hospital of Lima, Peru. Ludwig was experiencing severe pains and fever that were diagnosed as kidney problems.

Several years earlier, in 1965, Ludwig was traveling through Europe. His career as an international health food processor and distributor made him, as he says, "a perpetual globe-trotter." While in India, he met a normal-looking gentleman named Satu Ra.

After several weeks of friendship, Satu Ra told Ludwig that he, Satu, was actually an extraterrestrial. Although initially skeptical of such an outrageous claim, Ludwig was soon given reason to believe. Satu actually invited Ludwig aboard a UFO. Ludwig, of course, accepted.

Like most contactees, he was told many things. And like most contactees, Ludwig was on his own most of the time. Just because you have extraterrestrial friends does not mean that you can just sit back and let them do all the work. Contactees, like all people, must run their own lives.

But if trouble should rise, contactees are often given help. When Ludwig, for example, developed kidney problems, his alien friends were ready to help.

He was in the wilderness in Lima, Peru, when he developed his illness. Because of his extensive traveling experience, he knew how important it was to receive good medical care. He was very happy to get into the Maison Francais Hospital, which was well-known for its modern equipment and expert doctors.

The doctors told Ludwig that he would have to undergo an operation to alleviate him of his kidney pain. Ludwig agreed as his fever was getting worse and the pain had him totally incapacitated.

While in the hospital, Ludwig awoke one night to find a visitor in his room. He recognized the person as Xiti, Satu Ra's sister, whom he met aboard the UFO. Xiti wasted no time and immediately began treating Ludwig for his condition.

She began by passing her hands over his body. She then gave him a small pill to eat. Then, as quickly and mysteriously as she arrived, Xiti departed.

Within minutes after her administrations, Ludwig's condition made a complete reversal. He was instantly free of pain and his fever broke. As far as he could tell, Ludwig was completely healthy. The doctors were called in and were stunned. They realized that Ludwig was completely healthy, but could not explain why.

This case is well verified by the doctors, who knew that Ludwig should have been on the operating table. As Ludwig says, "All of this may be checked by any one of my readers. I became the miracle patient of the famous Maison Francais Hospital in Lima, Peru."

Ludwig had been scheduled for his operation that morning, so when the doctors examined him and found him perfectly healthy, they demanded he undergo a whole battery of tests. The tests verified the obvious, that Ludwig was perfectly healthy. As one doctor told him, "You're fit as a flea...this is really baffling. Yesterday you were a very sick man, and today you seem to be a different person."

Ludwig never had any more problems with his kidneys. He did have several more contacts with his alien friends and ended up writing a book about his experiences.[1]

One condition which can be extremely painful is known as kidney stones. Caused by calcified deposits, today's treatment usually involves ultrasonics which disintegrate kidney stones, allowing them to be excreted with the urine. Not surprisingly, extraterrestrials have also cured people of kidney stones, as this next case illustrates. What's unusual in this case is that the person who was healed was not a contactee, but rather, a friend of a contactee.

CASE #045. In August of 1980, Hector Vasquez of San Juan, Puerto Rico was awakened at 2:00 A.M. by severe side pains that had previously been diagnosed as kidney stones. Hector knew that he would have to go to the hospital, so he sent one of his children to get his friend, David Delmundo to drive him.

David is an ordained Baptist minister and also a contactee. After seeing many UFOs, he had a face-to-face encounter with a short, almond-eyed, jump-suited figure who said he was from the Orion Star Cluster. Thus began a complicated series of contacts.

Delmundo, however, had received no healings of his own, so he had no reason to expect his ET friends to assist his Earth friend. When he heard the news about the illness, he jumped up to drive Hector to the hospital.

On the way to the hospital, however, they had a close-up encounter with an "orange glowing luminous disk" which could be seen hovering ahead of them above the road. It then flew alongside the car, paralleling its course. Then it flew back in front of the car in a

---

1. Pallman, 1986, pp. 96-101.

strange, darting pattern. Vasquez was lying in the back seat, doubled up with pain, and when Delmundo said a UFO was hovering outside, Vasquez didn't believe him. Delmundo insisted Vasquez look, and to his surprise, Delmundo was right. By the way the object was moving, it was very clear to both Delmundo and Vasquez that the UFO was very interested in them.

Both of them wondered why the UFO was moving so strangely around the car. But by the time they had made the twenty minute drive to the hospital, they had their answer. Hector's crippling pain was completely gone.

Still, fearing another attack, he insisted upon a medical examination. The doctors performed a thorough examination and pronounced him in perfect health. His previously diagnosed kidney stone was nowhere to be found.

On the way home, the two men again saw the same object that had hovered around their car earlier. Hector Vasquez believes that he was healed of his kidney stone by the UFO. And, as of yet, Hector has not had a recurrence.[2]

Kidney stones today are easily treated. However, it wasn't always that way. Kidney stones used to entail a long and painful recovery, with a strong possibility of a recurrence. As these cases show, however, aliens have no problem when it comes to healing people of kidney stones. This next case is a perfect example.

CASE #038. In 1976, oil-field worker, Carl Higdon, of Rawlins, Wyoming, was out hunting in the wilderness of the Medicine Bow National Forest. To his delight, he saw five elk. He raised his gun, aimed it at a large buck, and fired. To his utter surprise, the bullet floated slowly out of his gun and dropped to the ground about fifty feet away.

At this point, Higdon noticed that he had a visitor. He saw a six-foot tall man, with yellowish skin, no neck, no visible ears, small eyes, a slit-like mouth, straw-colored hair and two antennas coming out of his head. Like many aliens, the man was dressed in a one-piece jumpsuit with a large belt.

The stranger gave Higdon some pills and encouraged him to swallow one, which Higdon did. His next memory is being inside a strange craft, strapped to a seat. A helmet was placed over his head,

---

2. Sanchez-Ocejo, 1982, pp. 158-159.

and strange images filled his mind. Higdon saw futuristic cities and other bizarre structures.

Next, Higdon underwent a physical examination. He also saw three other normal-looking people aboard the UFO, as well as several animals. Shortly later, Higdon was set back down. He radioed for help and blacked out.

Search parties were sent out. They found Higdon's truck mired in a field several miles from its original location. Also, some members of the party saw unexplained lights hovering high in the sky.

Higdon was finally found. Only under hypnosis was he able to recall what occurred aboard the UFO. His story has been told and retold in many UFO books. What is little-known, however, is that Higdon was ill before his encounter.

He suffered from a persistent and painful case of kidney stones. After the encounter, however, his problem with kidney stones was gone forever. He was completely healed. Higdon also received another remarkable cure of his lung, which will be recounted in a later chapter.[3]

The next case is one of the most amazing cases in all of ufology. Not only does it involve a well-verified UFO healing, but the healing occurred to a well-known UFO investigator. And even more astonishing, the UFO investigator knew he was going to have a contact. And when the contact occurred, it was actually witnessed by four other UFO investigators. It is truly a unique case.

CASE #019. On December 7, 1967, Danish UFO investigator Hans Lauritzen, and four of his friends were conducting an active field investigation of UFOs. Their research had led them to participate in routine sky watches in the hopes of observing a UFO close-up. Lauritzen had reason to believe that he might encounter a UFO in the near future.

One month earlier, Lauritzen had been contacted by a Swedish girl who told him that he was on the list of the Space People as one of the people to have a contact. However, sitting out in the field waiting for a contact was difficult. Lauritzen was suffering from an illness that made such work very tiring.

In February of 1966, Lauritzen had contracted a severe case of liver hepatitis and had to retire and go on a pension. At the time of the

---

3. Steiger, 1988, pp. 58-61.

UFO watch, Lauritzen was not only tired and weak, his liver was ten centimeters increased in size.

Just before ending the sky watch on December 7, the five men were surprised to see two "great, dim, yellow globes about a hundred yards away from them." Lauritzen immediately became entranced, walked away from the group and had a telepathic communication with the UFO occupants. The conversation was about helping humanity. They told Lauritzen that he has great power to help people.

After an hour, Lauritzen suddenly came out of his trance and heard his friends calling. He ran towards their voices and was surprised to find that he didn't even feel tired. As he says, "I ran and ran so fast that my four friends could not follow me. I had to wait for them. I realized that I had been cured of my hepatitis."

Lauritzen experienced some other disturbing symptoms, such as severe but temporary body pains, as well as strong emotional fluctuations and bizarre, destructive thoughts. All these were temporary, and today, Lauritzen feels that his encounter healed his liver and saved his life. As he says, "I passed a medical examination, not mentioning my contact, of course; and to the surprise of the doctors, ten centimeters of my liver had disappeared so that it was now normal size. The blood test showed that it functioned now as any other healthy liver...before I had a sick liver. At one time it was as much as sixteen centimeters too large. Now I have a healthy liver. I am most thankful to the UFO for having cured my otherwise chronic disease."

Lauritzen was able to return to work and resume his activities. He believes that this would never have happened without the UFO encounter. He is also adamant that other people believe him. As he says, "I swear to God that I have told the truth as far as I can see it."[4]

Another similar case of a liver healing took place in Finland. This case was also investigated by famed UFO investigator Brad Steiger.

CASE #069. A native of Finland suffered since birth from an abnormally enlarged liver. He had been told numerous times by doctors that his condition was chronic. Then, in the early 1990s, he was skiing down a mountain slope in a remote area when he was struck by a "white beam of light from an egg-shaped UFO." After the light left, the man realized he had missing time. He went to a doctor for an examination. It was then that he discovered the impossible; his liver

---

4. Steiger, 1992, pp. 145-146.

was healed. As the report reads, "During this examination, the doctors found that his liver had been reduced to a normal size."[5]

Yet another case of a liver healing comes from UFO investigator, Edith Fiore, Ph.D. Fiore is directly responsible for uncovering many dramatic healings of all types. The following case is no exception.

CASE #099. Abductee Linda X. of California has received several cures at the hands of aliens. One of her many cures was for a persistent liver infection.

As a child, Linda fell down and damaged her liver. This left the organ very weak and vulnerable to dangerous infections. As a child, this caused her many health problems. Unknown to her at the time, she was being regularly treated by the extraterrestrials.

Under hypnosis, she was surprised to find out that they had been caring and administering medicine to her for a long time. As she says, "...over the years they have been slowly refurbishing, just as if building, like building it up. The salve penetrates the whole body and generates new cells that will fight the infection and in turn will give off new life in my body."

Linda's cures are among the most astounding ever recorded. As we shall read in a later chapter, she was also cured of cancer by the aliens.[6]

As these accounts show, the kidney and liver cures have been reported by many different people who have no knowledge of each other's accounts. And yet, the stories are remarkably consistent. Over and over again, people report sudden and total reversals of their illnesses after a close encounter with a UFO.

The conclusion is inescapable—UFOs are actively healing people's livers and kidneys across the world. The majority of the witnesses in these cases were examined by physicians who pronounced them healthy. In some of these cases, the doctors were able to examine the patient before and after the cure, thereby increasing their credibility.

For those who believe that one or another of these cures were due to misdiagnosis, hoaxes or natural healing, there are always more cases to challenge their skepticism. In the next chapter, we shall examine another group of healings involving another area of the human body—the lungs.

---

5. Steiger and Steiger, 1994, p. 40.
6. Fiore, 1989, pp. 102-103.

KESARA

# Lung Cures

The lungs are among the most vital of all organs. The number of diseases and illnesses that affect this organ are almost endless. Moist lungs provide the perfect breeding ground for airborne bacteria and are therefore susceptible to all types of infections such as pneumonia, bronchitis or even tuberculosis.

One condition which afflicts many people can also be very deadly. The condition, known as asthma, actually kills people every year. An asthmatic crisis is a spasm and restriction of the bronchial tubes that transport the oxygenated blood. There are some indications that it may be partially psychosomatic as some cures have been effected through hypnosis, however this remains a controversial issue. Unfortunately, the majority of asthmatics must suffer with their sometimes crippling disability.

The ability to breathe easily is something most of us take for granted. Only the person who has suffered from asthma can truly know how it feels to not be able to get enough air. To remedy their condition, asthmatics usually carry around a portable bottle of medicine such as a bronchial spray.

There are some people, however, who have another safeguard. In the following case, a young girl received treatment for her asthma by extraterrestrials from Andromeda.

CASE #063. Bert and Denise Twiggs of Hubbard, Oregon, have figured prominently in this book. As contactees, they and their children have received a number of cures.

The Twiggs' daughter, Stacey, suffers from severe asthma. On October 15, 1989, during the evening, Stacey's asthma was acting up. In the middle of the night, she was taken onboard a UFO and treated. The aliens explained that the asthma was caused mostly by pollution and that the only real cure would be clean air. They also gave her "some medicine to help her through her current attack."

In January of 1990, Stacey suffered another severe asthma attack. When on board the UFO that evening, the alien doctors told the Twiggses that there was little that could be done and that the attacks would get worse.

The next day, Stacey's lung capacity was reduced to a dangerous twenty percent. In fact, by the time Stacey made it to the hospital, one lung was in the process of collapsing. The human doctor simply confirmed what the alien doctors had already told the Twiggses. There was little that could be done, except to keep her on medication.

Still, the alien doctors did what they could. As Denise Twiggs says, "...the Androme doctor saw her each night. He was monitoring the Earth doctor's orders. He also added some of his own medication to her recovery."[1]

Another cure of a more dramatic nature took place in England to a man who had no reason to expect such a cure. The man was not a typical contactee. He had only seen a UFO once, and then briefly. Four years after the man's encounter, he developed a serious lung condition. Evidently, the aliens were aware of this; they took steps to cure the gentleman of his condition.

CASE #018. On Easter weekend in 1962, electronics specialist Fred White went fishing near Durhan, England. While fishing, he experienced a close-up sighting of a metallic craft with portholes. Through the portholes, Fred could clearly see a human-looking person. After about five minutes, the UFO left.

Four years later, in September of 1966, Fred was hospitalized because of severe chest pains and a collapsed lung. Fred had no idea what was wrong with him. To his horror, X rays revealed that he had a large hole in his lung.

During his stay in the hospital, Fred was visited by a man whom he thought at first was just another doctor. However, the doctor was strange. He had a peculiar accent, and when he entered the room, he immediately put his arms over Fred as if examining him. Fred noticed that the man wore "some kind of watch which glowed like no timepiece I had ever seen."

---

1. Twiggs, 1995, pp. 112-113, 131-132.

The two of them began discussing Fred's symptoms. The man also expressed some interest in Fred's work with electronics. After several minutes, the strange doctor left.

Fred reports that, as soon as the stranger departed, he felt immediately better and could breathe without pain. The doctors gathered around and demanded more tests. X rays were again taken. To the doctors' surprise, no hole in the lung could be found, and Fred was given a clean bill of health.

One astonished doctor told Fred, "Your lung is completely inflated. The hole you had in there is healed. We've never seen anything like it."

Fred White is completely convinced that the man who visited him in the hospital cured him of his illness. He also has a strong suspicion that the man is the same person he saw inside the UFO while fishing, four years earlier.[2]

While most cures by the aliens seem to be purposeful, there are some that seem to be coincidental. It seems that by simply being inside or even near a UFO, miraculous healings of all types take place. The following case is a perfect example. It involves a man who suffered an abduction, only to find out much later that he had also received multiple cures. The case is a very well-known abduction case. Very few people, however, are aware of the healings which took place during the encounter.

CASE #039. In 1976, oil-field worker Carl Higdon of Rawlins, Wyoming, was abducted aboard a UFO.

Before the encounter, Higdon suffered from kidney stones and also had a tubercular-like scar on his lung. Higdon's problems with his kidney stones disappeared immediately. It wasn't until a couple of years later that he was told by his doctor that the tubercular scar on his lung was gone. Higdon feels that his lung was healed while he was aboard the UFO.[3]

Although the lungs are susceptible to a number of serious infections, one of the deadliest is known as pneumonia. Pneumonia is particularly dangerous because it can easily creep up on someone. A person might have a cold one day, and find themselves on their death bed the next. Pneumonia kills large numbers of people every year. It

---

2. Johnson, July 1988, pp. 26-27.
3. Steiger, 1988, pp. 58-61.

is basically a bacterial infection which causes the lung sacs to fill with pus. Fortunately, most pneumonias respond quickly to antibiotic treatments. Nevertheless, the disease leaves millions hospitalized and often kills the old and weak.

The following case is a good example of how pneumonia can sneak up on a person, and convert an annoying condition into a fatal condition in less than a day.

CASE #041. In July of 1978, world-famous contactee Eduard "Billy" Meier of Switzerland came down with a bad cold that developed almost instantly into severe pneumonia. Eduard was so weak, he was scarcely able to hold up a glass of water to drink.

Meier, however, happens to be one of the most influential contactees to date. His case is highly controversial because of the large number of crystal-clear photographs of UFOs taken by Meier. Despite this, or perhaps because of it, the case has divided the UFO community.

Meier, however, has a mission to perform, which is in part to tell the world about his contacts. Evidently he is succeeding as there are over eight books and five films about the case, with more of each on the way. Because the Meier case seems to be of such profound importance, it is not at all surprising that when he developed pneumonia, the aliens were there to help.

Meier was given instructions to go to a certain location. He had his friends take him to the spot. Evidently Meier's condition was life-threatening for the aliens boldly took Meier in full view of his friends. Several witnesses clearly saw Meier disappear in front of their eyes in a "flash of bluish light."

Meier says that he was taken inside the UFO where he underwent an operation. Human-looking aliens used "strange instruments" to drain the pleural area of his chest of "a considerable amount of pus and liquid." His alien friends told him in no uncertain terms that if they had not intervened, he would surely have died. They told him that next time, he should not delay to see a human doctor.

A half-hour later, he was let out of the craft and rejoined his astonished friends. The change in Meier left his friends speechless. Before being taken aboard, Meier was near death. His face had a "gray color" and his eyes had a "deep sunken appearance." Afterwards,

Meier's pneumonia was gone. He was "sprightly and smiling," and appeared to be "in perfect health."[4]

Pneumonia cures may be actually more common than many people suppose. David Jacobs, author of *Secret Life*, is one of the world's leading authorities on UFO abductions. Although Jacobs is somewhat skeptical of many contactee claims, he does admit that some of his abductee clients report being cured by the extraterrestrials. As he says, "At least two abductees have reported that their cases of pneumonia were cured during their abductions."[5]

As it turns out, Eduard Meier is probably the only person on record who has been cured twice of pneumonia by the aliens. His first cure occurred when he was an infant, as the following case shows. It is the second earliest recorded case of a UFO healing.

CASE #002. In 1937, when contactee Eduard Meier was only six months old, he contacted pneumonia. He learned later on that the family physician, Dr. Strebel, expected Meier would die. He told Meier's parents that nothing could be done and that they should prepare themselves for the death of their son.

That evening, Meier had his first extraterrestrial contact, with an alien named Sfath. Sfath administered to Meier and cured him of his pneumonia.

Meier was told of this incident at a later contact, so he asked his mother to verify the information. To his surprise, she said that it was true; he had nearly died of pneumonia as a baby, but had suddenly and inexplicably recovered. His mother remembered how they had spoken of his recovery as a "miracle."[6]

The following case also involves a cure of pneumonia. This case was investigated by John Mack M. D., and appears in his book, *Abduction*.

CASE #003. In 1942, Edward Carlos (age five) of Philadelphia, Pennsylvania, developed a severe case of respiratory pneumonia. He had a high fever and was in a near coma. The illness was too much for Carlos to bear, and he had an out-of-body experience. Once outside his body, he found himself surrounded by grey-type aliens.

---

4. Stevens, 1982, pp. 240-241.
5. Jacobs, 1992, p. 191.
6. Stevens, 1982, pp. 42-43.

His astral body was laid down and beams of light were shone upon it. Carlos describes the lights as being "like laser beams coming into my body through the soles of the feet and the hands, and possibly through the sides of the lower torso, radiating throughout the whole body, expanding and changing color as the light grew to fit the whole body interior, thereby healing it."

According to Carlos, the light was yellow in the middle and surrounded by successive concentric layers of red, blue and green. After the procedure was finished, Carlos was returned to his body. His fever was broken and he had come out of the coma. Carlos has also received other cures.[7]

As we have seen, cures of lung diseases are quite numerous. The above cases represent the typical cross-section of lung cures as performed by extraterrestrials. Although perhaps not perfect, alien medical technology is obviously much more advanced than human medical technology.

Up to now we have examined illnesses, injuries and conditions from which most people recover. In the following chapters, however, we shall cover diseases which leave doctors baffled and even downright scared.

---

7. Mack, 1994, pp. 334-344.

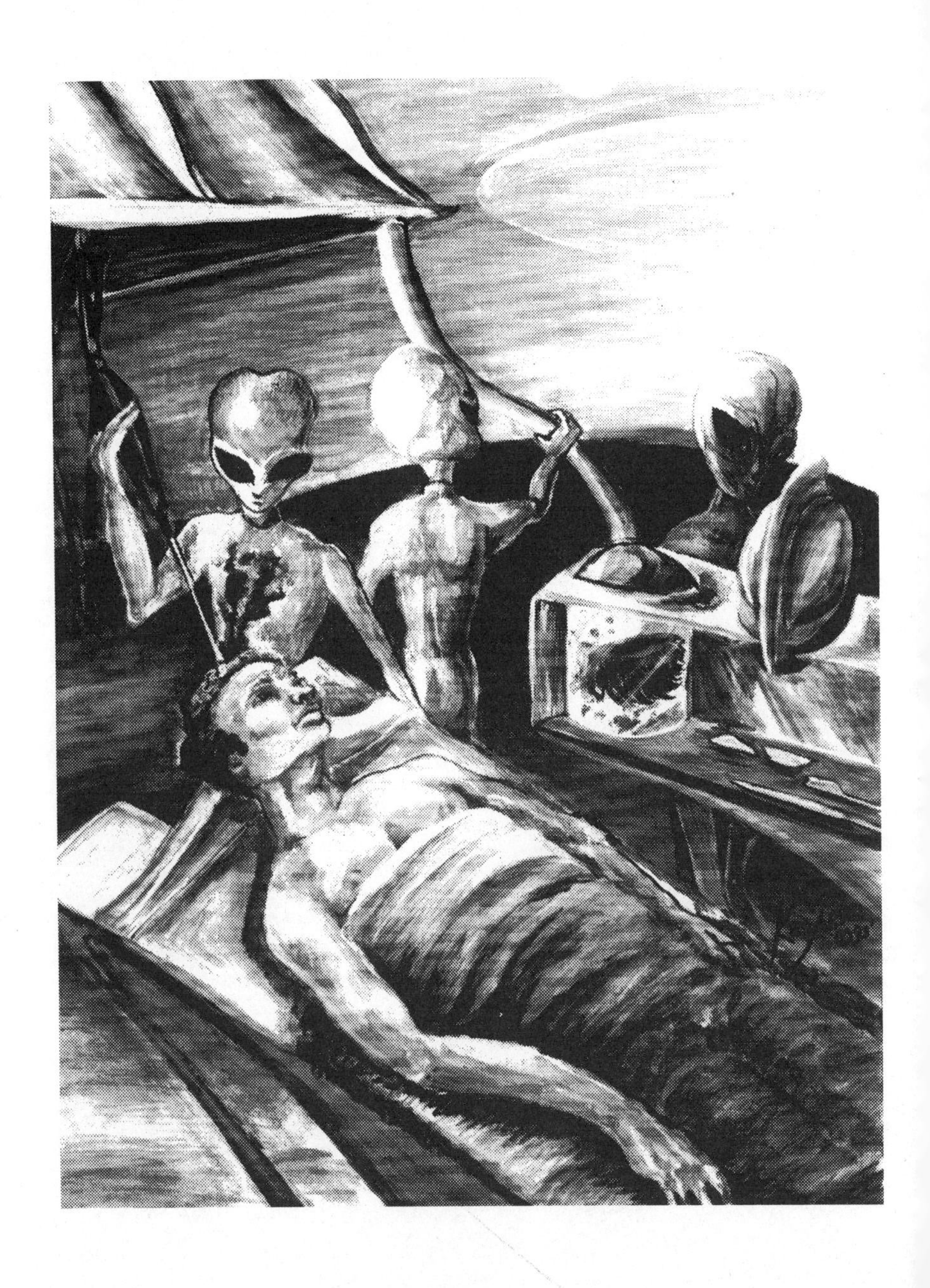

# 9

## Healings of Serious Illnesses and Chronic Diseases

Disease has always been one of the most feared and mysterious things on Earth. The evolution of disease on Earth is fascinating. Throughout history, various plagues and epidemics have wiped out large portions of humanity. Diseases like the black death (bubonic plague), cholera, polio, diphtheria, typhoid, small pox and countless others raged unchecked through the population, killing millions of people.

It wasn't until the germ theory of disease origins became popular that modern human doctors exercised some control over the unseen killers. Then came the widespread use of vaccinations. Soon the medical profession had access to penicillin and many other wonder drugs.

Diseases, it seemed, had finally come under control.

Of course, this was a major step forward, but hardly the end of disease. Many diseases continue to exist whose causes are still only partly understood.

Consider the disease known as multiple sclerosis, which is a chronic disease of the nervous system. The symptoms include shaking limbs and uncontrolled spasms, difficulties with speech and eye movement, and an unsteady gait. There is no known cure, nor are the causes understood.

However, the following case of a UFO healing involves a curing of this dreaded disease. As we shall see, the healing of this so-called "chronic" disease seems to have been permanent.

CASE #042. On August 30, 1979, retired Canadian Army officer Jean Cyr of St. Eustache, Quebec, and his family were drawn outside by a strange humming sound. Once outside, they saw a large, glowing, metallic disk rise from the fields behind their house and hover over their home, flashing down bright beams of light. They called up the nearby Mirabel Control Tower and the Montreal International

Airport, who also confirmed that they could see the UFO. The tower controller vectored in a plane, but the UFO winked out and came back when the plane left. Another attempt provoked the same response. Meanwhile, neighbors of the Cyrs saw the object and called in the police. Five hours later, the UFO left, leaving strange circular marks in a nearby field from where it was initially seen rising.

Mr. Cyr and his family were extensively investigated by various UFO groups. Meanwhile, over the next few weeks, they began having more sightings. During one sighting, the UFO was again observed hovering low over the Cyr residence. Jean Cyr was able to get an audio recording of the UFO. Many other people in the area also saw the UFO. Cyr also noticed strange marks on his body that couldn't be explained.

Although this case has many fascinating aspects, the most fascinating of all is the amazing cure received by Jean Cyr. Before the encounter, Cyr suffered badly from multiple sclerosis and had to go to the hospital for treatment every two weeks. Normally, he would be in a wheelchair virtually paralyzed. However, after the series of encounters, Cyr appeared to be completely cured. As the report on the case reads, "...all symptoms of the disease have gone, and he hasn't had a stroke since."[1]

As can be seen, diseases of the nervous system are quite serious indeed. However, the nervous system can be destroyed in ways other than disease. In the following case, a young man suffered from a severe cerebral aneurysm which completely destroyed his nervous system.

Fortunately for him, he was the close friend of a contactee. And by the good graces of the aliens involved, a cure was affected.

CASE #050. In 1984, Joao Vicente of Botucatu, Brazil, suffered a near-fatal cerebral aneurysm. He was rushed to the hospital and put on life-support equipment. Unfortunately, his aneurysm was so severe that the brain was flooded with blood, destroying the nervous system. The pressure inside Joao's skull was so heavy that surgery was impossible. Doctors recommended removal of the life-support equipment because "there was nothing that modern medicine could do for him."

---

1. Stevens, 1982, pp. 451-455.

Then came an interesting turn of events. Joao Valerio, the hospital doorman, was a close friend of Vicente. Valerio also happened to be in contact with extraterrestrials. Valerio had experienced numerous contacts, was taken on tours of the solar system, was given predictions and physical artifacts and was also permitted to take photographs of the UFOs and the aliens on their home planet.

Valerio was well-known for his contacts, so the doctors in the hospital asked Valerio if he would ask the aliens for some medicine. Valerio agreed and during his next contact, was given medicine by the ETs to give Vicente.

UFO investigator Rodolfo R. Casellato was one of the principal investigators of the case. Permission has been graciously given to quote extensively from his report, which is published in the book, *UFO Abduction at Botucatu*, published by the UFO Photo Archives. As Casellato says, "Joao [Valerio] obtained from the ETs a substance that after a simple analysis showed a very low fusion point. It melts at body temperature…the instructions were that the substance was to be rubbed into the foot sole, and something would form on that place and the residue from the brain (hemorrhage) would come out there.

"And that is exactly what happened," Casellato continues, "and Joao Vicente started getting a little better. Sometime later, Joao Valerio brought another substance to be rubbed on the spinal column, and that is being done now…X rays are showing now that a new nervous system is taking over the older one destroyed, new arteries, and the damaged organs are being encapsulated by new tissue, all shown in the X rays. The doctors call it a 'miracle'…I saw Joao Vicente and he shook my hand and spoke some words and is getting better."

This cure is not only well verified, but investigators managed to procure some of each batch of medicine for chemical analysis.

Neither substance was identifiable. As the report of the first medicine reads, "The substance was in soft crystal form with the crystals about the size and color of raw sugar. The crystals had to be kept away from heat because they began to melt down at about 115 degrees Fahrenheit. When held in palms and rubbed together, they decomposed into a greasy substance that was then absorbed or evaporated away or both. The result of the testing was that the substance was "chemically unique."

The second substance was never analyzed due to lack of funds; it costs over one thousand dollars for a typical analysis. The second substance, which was applied to Vicente's spinal column, was described as "a dark irregular granular material from very fine up to 3 mm across. It was not in true crystalline form, as it lacked regularity, but looked more like it had been crushed, more like small shards. These pieces were semi-transparent, and of a dark reddish-brown color like shattered garnet stones."

The full account of this amazing case, and the above mentioned photos are all contained in the book, *UFO Abduction at Botucatu*.[2]

The extraterrestrials' ability to cure disease seems nothing short of miraculous. Our human doctors are often helpless in the face of such horrendous health problems. However, the alien doctors seem to have no problem in this area. Their ability to "regrow" a new organ system may seem like science-fiction, but more than one person has reported this type of cure.

Consider the following case from UFO investigator Edith Fiore, Ph.D., in which a young boy was cured of an extremely rare disease that would almost certainly have proved fatal. This case also illustrates how shockingly advanced the aliens are in the field of medicine.

CASE #080. At the tender age of eighteen months, "Ted" of Santa Clara, California was diagnosed with a rare congenital disorder known as angioma, which is a vascular malformation in the brain. Typically, this condition puts tremendous pressure on the brain, usually causing epilepsy. In Ted's case, the blood vessels in his brain were abnormally large, causing a variety of devastating symptoms. His case was so severe that doctors gave him only a few years to live.

Meanwhile, Ted's condition rapidly deteriorated. He became emotionally unstable and suffered intermittent paralysis. His vision weakened until he had only double vision. He lost his ability to walk and was forced to crawl. By the time he was three years old, he could no longer talk or feed himself, and the right side of his body had simply stopped growing.

About this time, when Ted should have died, his condition began to miraculously improve. His parents took him to the hospital for exploratory surgery. The surgeons, to their amazement, noted that the tissue growth had not only stopped, but shrunk. They were unable to provide an explanation.

---

2. Casellato, da Silva and Stevens, 1985, pp. 37-39, 194-220.

Ted, however, had an explanation. He clearly remembered going aboard a UFO and being cured of his condition. He was placed on a table where an apparatus was placed over his head. Several penlike instruments were pointed at his skull. The instruments emitted beams of light which entered inside Ted's head, curing him of his condition. As Ted says, "When they turn the apparatus on, a beam of light comes out. It's very straight, and it's very thin. It's needle-like. It's very much like a laser light. And it comes out and strikes my head, but it doesn't hurt."

Ted also reports that the aliens cut open his skull with another laser and examined his brain. Using the healing laser lights, the aliens began to probe his opened brain. As Ted says under hypnosis, "...they touch all over and they find enlargement. And with the laser light, they're able to shrink it."

Ted reports that the light had the "thickness of a spiderweb" and seemed to be "almost like silk." After the beams were shone on his brain, another suction-like instrument was placed over his brain, and an ashlike substance caused by the laser was removed. The top of his skull was then placed back over his brain and another laser was shone on the wound. And like many witnesses record, the cut was healed by the light. As Ted says, "And this time instead of cutting, it mends. It mends the skull back together. The atoms go back together and there's no scar tissue. And when it's finished, there's no indication that the skull has been opened."

Needless to say, Ted's cure was permanent.[3]

Despite the many advances of modern medical science, a large number of killer diseases continue to plague humanity. One disease that has resisted all attempts to eradicate it is known as tuberculosis.

This infectious disease continues to spread throughout the population. Tuberculosis can be prevented with inoculations and can usually be cured with antibiotics or by the body's natural immune defenses.

In some cases, however the disease may lie latent for months or even years, showing no visible signs. Some people can become carriers of the disease and yet, never show symptoms. The symptoms, if and when they do appear, include night sweats, weight loss, fever, coughing and spitting up blood.

---

3. Fiore, 1989, pp. 132-145.

If left untreated, the disease can lead to a number of health complications and even death. In the following case, death from tuberculosis would probably have occurred had not the extraterrestrials intervened. This cure is the earliest UFO cure on record.

CASE #001. In the mid-1930s, an anonymous gentleman, today a famous broadcaster in his seventies, reported that he contracted tuberculosis at the tender age of thirteen years. At the time, there was no such thing as vaccinations, inoculations or even antibiotics. Diseases such as tuberculosis simply could not be controlled. As the gentleman says, "I got sicker and sicker, and the doctors thought I would die."

While on his deathbed, the gentleman had a "dream" in which he was taken aboard a spaceship and cured of his disease. As he says, "In the 'dream,' a spaceship took me aboard and placed me on a table within a glass bubble...the creatures did medical procedures on me and I was made to understand that now I would be well and that for the rest of my life I would be healthier and stronger than normal humans. When the doctor saw me a few days after the dream, he found no trace of the tuberculosis."

Despite the miraculous cure, the gentleman was skeptical that he had really been on board an actual UFO. In those days, stories of spacemen were very rare. It wouldn't be until the fifties when such stories became more popular that the gentleman began to recognize what happened to him. As he says, "I began to hear other people recount incidents of being taken aboard alien craft, and their description are much the same as mine right down to the way the creatures looked."

Incidentally, the gentleman claims that in the late 1980s, his car was hit by a train, totally destroying it. He was inside the car at the time and didn't receive a scratch. It is his firm belief that the aliens saved his life. Reportedly, railroad investigators also believed that the accident should have been fatal.[4]

Another serious disease that affects many people is known as muscular dystrophy. The causes are unknown but it is known to be a genetic disorder. The symptoms include the weakness and degeneration of muscle tissue. Eventually the muscles are replaced by fatty tissues.

---

4. Overstreet, Oct. 27, 1992.

Although there are several treatments for the disease, according to official medical science, it cannot be cured. It is a chronic condition.

In the following case, however, the disease was not only cured, but was done so in a matter of seconds! Again, the alien doctors put most human doctors to shame.

CASE #043. Dean Anderson of Sturgeon Bay, Wisconsin, came forth with his story of his 1979 voluntary visit aboard a UFO. He has several friends who also claim to have contacts with the same friendly aliens. Dean says that he has been aboard UFOs many times. His alien friends appear to be humans but say that they are from Saturn and Jupiter.

While visiting aboard the UFO, Dean observed many things. Pertinent here is the healing he saw take place. According to Dean, during one contact in 1979, he saw a nine-year-old girl and her mother, both from Detroit, Michigan, aboard the UFO. He was told that the girl was suffering from muscular dystrophy.

Dean watched with fascination as the girl was placed on a table with a "glass shield" over her body. She was then "hooked up to an apparatus which caused the table to be suspended in mid-air; for twenty seconds everything glowed with a brilliant blue-white aura. The table settled, two short fellows came in, lifted the girl off the table, and she walked over to her mother—perfectly healed. I was amazed!"[5]

As hard as it is to believe, serious diseases seem to pose no threat to the aliens. Take the disease known as diphtheria. Before the age of vaccinations, diphtheria was a dreaded disease which killed large numbers of young children. The illness is caused by the bacterium *Corynebacterium Diptheriae,* and starts with a sore throat, weakness and mild fever.

At this point, a soft grayish membrane forms in the throat, causing difficulty in breathing and swallowing. Soon, the bacteria reaches toxic levels which can lead to permanent damage to the heart and nervous system. The disease can kill in as little as four days.

Fortunately, the disease responds to antibiotics and today, most children in the United States are immunized against the disease by vaccination.

---

5. Beckley, July 1988, p. 6.

Still, however, there are isolated cases, which sometimes prove to be fatal. The following case involves a severe case of diphtheria which was probably fatal, except for the fact that extraterrestrials intervened and cured the victim of the disease.

CASE #016. In 1962, Alice Haggerty, then six years old, was stricken with diphtheria. Because of a religious background, Alice had not been vaccinated against the disease. For the same religious reasons, her parents refused to take her to the hospital. The doctor said that the condition was serious, but Alice's parents refused to listen.

Over the next two weeks, her condition continued to deteriorate. The doctor visited daily, and at the end of the two weeks, he told Alice's mother that he did not expect Alice to live through the night.

Alice, however, happened to be an abductee, and would eventually have several abductions. She tells the complete story of her abductions in an audio-taped interview available from the International UFO Library Magazine. She is also one of the cases presented in David Jacob's book, *Secret Life*, under the pseudonym, Lynn Miller. She has released her real name in her audio-taped interview.

On the evening in question, Alice was abducted by friendly aliens which she then described as "angels in white robes with silver belts." Alice remembers being transported inside a UFO where the aliens told her she would be cured. They waved a small "rodlike device" around her body, and then made her stand inside a "large blue cylinder" with a little window. The aliens watched as a bright blue light was emitted from the top into the center of the cylinder. The light moved slowly down from the top of the machine until it was about eight inches from the top of Alice's head. The light then retreated and Alice was instructed to exit the cylinder.

According to Jacobs, the aliens "informed her in a matter-of-fact manner that she was now cured and 'cleansed.'" Alice then underwent more of the typical types of procedures that are often performed aboard UFOs. The next morning, Alice's mother found her playing on the floor as if everything were perfectly normal.

Alice was ordered back into bed by her mother and her doctor, but both admitted that the diphtheria was completely and totally gone.[6]

---

6. Jacobs, 1992, pp. 191-192, and Haggerty, 1991, p. 34.

Another serious and chronic disease that afflicts a great many people is rheumatism. Rheumatism typically causes various joints in the body to become stiff and inflamed. The condition is very painful and, although treatments are available, there is no known cure.

However, the fact that there is no known cure does not mean that a cure is non-existent. In fact, the following case is a solid example of someone who was cured of rheumatism by a UFO.

CASE #023. On December 9, 1968, an unidentified Peruvian Customs official was on his deck when he was struck on the front side of his body by "violet rays" from a UFO. The customs official was instantly cured of two disorders. One was the dramatic improvement of his vision. Before he was severely myopic and wore thick glasses. After the encounter, his vision was perfect.

In addition, he suffered from rheumatism, that is, until his UFO encounter. After being struck by the violet rays, his rheumatism completely disappeared.[7]

Another serious disease which affects millions is arthritis. Arthritis is basically an inflammation of the joints. In most cases, it is a chronic disease. It can also be painful and crippling. It can strike anyone at any age, but is most common among the elderly. In the following two accounts, people were cured of crippling arthritis after a close encounter with a UFO.

CASE #057. Anne DeSoto of Watsonville, California, was driving with her boyfriend one afternoon in December, 1988 when their car was suddenly enveloped in fog. Shortly later, they saw a bright, orange globe of light hovering over a nearby field. DeSoto had seen UFOs before so they turned their car and headed towards the light.

At that point, they experienced a period of missing time. Their next memory is walking into DeSoto's boyfriend's house nearly five hours later. They both went inside and fell into a deep sleep.

The next morning, they both found that the bedsheets had blood stains. Anne also had a strange-looking mark on her hip. Even more amazing is that she was cured. As the report says, "...the arthritis she carried for years in that hip, disappeared."[8]

---

7. Blum and Blum, 1974, p. 147.
8. Townsend, Nov. 11, 1993.

CASE #065. Mae X. is a sixty-year-old waitress employed at a truckstop in Mississippi. Her job required her to be on the road in the early morning hours. On December 15, 1989, at four A.M., Mae was heading home when a bright light hovered over her car. Suddenly, it was six A.M., and Mae found herself waking up with her car parked along the side of the road. She had vague memories of being in a strange "doctor's examining room." She remembered several figures standing around her while a "doctor" stood next to her and "passed some kind of rod over her body again and again."

Mae was perplexed as to what happened to her, but she was even more surprised to find that she had physical symptoms of the encounter. Mae had suffered from severe arthritis for several years. In fact, the arthritis was so bad, that Mae found that she was about to be forced into retirement with no benefits. The arthritis was present in her wrist, knee and finger joints. After the missing-time abduction, Mae discovered that her arthritis was gone. She no longer felt any pain, allowing her to continue her employment.[9]

The number one killer of Americans today is heart disease. There are at lease five cases on record of healings involving the heart. The first case comes from Katharina Wilson, who has written a book about her UFO experiences titled, *Alien Jigsaw*.

CASE #060. On August 7, 1989, real-estate broker Katharina Wilson of Portland, Oregon, was injured by a nearby strike of lightning while staying in Pensacola, Florida. She knew she was injured by the strike, but decided to wait until morning to see if her condition improved. Katharina also happened to be a UFO abductee, having had a huge number of encounters. That night, she was visited again.

She woke up to find herself lying on her back on a table, suffering from excruciating chest pains. As Katharina watched from a dissociated state, grey-type aliens cut a square into her chest and attached a black mechanism with several extensions to the hole in her chest.

When the pain became particularly intense, the aliens telepathically told her, "We are repairing your heart. You will be okay now."

The next morning, Katharina woke up with a sore chest, but was otherwise healthy. As she says, "The first thing I did when I got out of bed was to look for a scar or a square cut into my chest. I found nothing. No blood on my sheets and no scar. My chest was sore

9. Steiger and Steiger, 1994, p. 41.

throughout the day, but it was not as sore as I would have expected it to feel after such a radical operation. I believe this machine they had over my heart was realigning the electrochemical impulses in my heart because they had been altered by the lightning...somehow I believed the aliens were repairing the damage the high voltage of the lightning had done to my heart."

Katharina reports that she has undergone a large number of operations at the hands of aliens. As she says, "They have performed surgery on me many, many times! ...It was shocking to remember the different times the aliens had performed surgery on me...What could the aliens be doing to me that would require my having surgery so often? It was not the first time I felt I may be a part of a huge experiment."[10]

Arterialsclerosis is one of the leading forms of heart disease. Fortunately, we have many new treatments of heart disease. One of these treatments is radical but effective. It involves inserting miniature lasers into major arteries to clear them of fatty obstructions.

In the following case, extraterrestrials used this exact procedure.

CASE #068. Edward Carlos, a fine arts professor in Pennsylvania, has had many UFO encounters. In fact, he is one of the clients of John Mack, M. D., and is featured in Mack's book, *Encounters*. On April 15, 1990, he was visiting the island of Iona between Scotland and Ireland when he saw an inexplicable beam of light coming down from the clouds. Shortly later, the beam struck him and he experienced a missing time abduction. In the past, he had had many UFO encounters, some of which he remembered consciously. He began seeing Mack and undergoing hypnotic regression.

He was able to recover the memories from the 1990 incident under hypnosis during which he recalled being examined by several entities, including a reptilian-like humanoid, small white-skinned humanoids, and insect-like robotic creatures. He recalled having an anal examination to determine his health. As he says, "They are clarifying that the inside of me is okay, and they are operating on me...If there is anything not right, the process can be healing."

As he lay there, they examined his various organ systems, including his heart. They then used a laser-like light to clear Carlos' arteries. As he says, "They can cause the change with their laser-like

10. Wilson, 1993, pp. 127-129, 250.

instruments...In my heart, I feel an extreme heat...I think it is healing something. Clearing arteries or something."

Carlos reports that his abductions, although they often involve healing procedures (including a possible cancer cure), have ironically left him with physical scars, bumps and rashes. Still, he feels that his encounters are positive and transformative. As he says, "You are diseased and then you are healed. With each healing, the emotional growth is established and connected in the human realm and I can go and utilize that towards teaching others."[11]

This next case was investigated by Barbara Lamb and remains one of the most extraordinary healings on record, involving the healing of a congenital heart defect. The first healing was performed in utero, before the witness was actually born. Then there were follow-up healings throughout the witness's life!

CASE #005. An anonymous male from southern California was born in the early 1940s with a congenital heart defect. According to the his doctors, the man should not be alive. As Barbara Lamb reports, "The first one is a man I work with fairly frequently who was born with a hole in his heart. And that was apparently taken care of by one particular extraterrestrial even beginning with when he was in the womb...The initial healing according to our regression started when he was in his mother's womb. He said, here he was in the womb and there was all this warmth coming through to him, and he was aware that this one particular extraterrestrial who has followed him all along, was placing his hands on the mother's womb—she was probably asleep I would guess—and then added or kind of sent this healing energy. And the extraterrestrial said at that time that he was healing the hole in the heart with an invisible electromagnetic strip...There was some healing done there. We've got that in a regression.

"And then we did another regression where they were doing additional healing when he was age seven. And he was taken aboard the spacecraft. Actually, that would have been a fatal thing to have a hole in your heart, but he did live. And he lived all those years until he was forty-eight and finally had surgery on his heart. And when he had surgery on his heart, the doctors expressed amazement that he had lived with that, which was obviously an old congenital defect. It

11. Mack, 1994, pp. 348-368.

was very hard for them to understand why he had lived. They were amazed...the doctor said, 'This is amazing. It looks like something has been somehow holding this defect together. But we can't figure out what it is.''

At the time, the witness had no memories of any healing extraterrestrial encounters. He had sought out Barbara Lamb after having other UFO experiences. He had no idea that the aliens had been healing him throughout his life. Today, he remembers many experiences with the aliens which he describes as five foot tall, grey-type aliens with eyes that are bevelled or multi-faceted like a diamond. He reports that the aliens have been teaching him how their technology and how to pilot their ships.[12]

The following report comes from Michael Hessemann, a well known and highly respected German UFO researcher. He lectured on the case at the 1994 UFO Expo West, held in Los Angeles, California. The case involves a man who was cured of a heart condition.

CASE #087. An anonymous factory worker from Georgia, Russia, had recently suffered two consecutive heart attacks. As any doctor will tell you, the chances of surviving a third heart attack are not very good.

One evening, the man woke up with severe chest pains and shortness of breath. It was another heart attack. Before he was able to take any action, he found himself surrounded by the typically described grey-type aliens. The aliens took a "ball of light" and put it inside the man's chest. He was then telepathically told that he should paint his experience.

He unaccountably fell asleep and woke up the next morning healed. He immediately began taking steps to make a painting of the creatures. Today, he is perfectly healthy and runs daily marathons in Gorky Park.

Hessemann reports that he has investigated several healing accounts in this area of Russia. Unlike the United States which abounds with frightening abduction accounts, Hessemann reports that in Georgia, Russian people are often eager to be taken aboard.[13]

Another report, though sparse on details, is still worth mention as it shows how widespread UFO healings are.

---

12. Personal files.
13. Hessemann, Michael. Lecture at 1994 UFO Expo West, Los Angeles, CA.

CASE #094. In Tblisi, Georgia, Russian people are allegedly having contacts on a daily basis. This information comes from UFO researcher Helga Morrow who owns a rare Russian UFO documentary. Says Helga, "After almost six hours of studying this film carefully, I feel it is the most outstanding first-hand information about extraterrestrials ever to be filmed anywhere in the world."

The film recounts several healings, including the following: "The shaven stomach of a man shows a slight scar. He was permanently and painlessly cured of a heart condition." That is the sum total of the report.

Cures from this area have appeared earlier in this book, and as we shall see in a later chapter, it is the location of several other dramatic cures by extraterrestrials.[14]

Yet another disease that causes much needless suffering is diabetes. Diabetes is a metabolic disorder which often becomes apparent when the patient exhibits excessive thirst and urination. The most common type, diabetes mellitus, is caused by the body's failure to produce enough insulin, which helps to regulate the blood-sugar level. Diabetes, if left untreated, can quickly lead to coma and death.

Unfortunately, there is no known cure. There is, however, an unknown cure, as this next dramatic case clearly shows. Unfortunately, there is no date provided for the case.

CASE #095. In Tblisi, Georgia, a young diabetic girl told her family that she had been contacted by friendly aliens who cured her of her disease. According to her account, she was awakened by an alien with the name of Bonytari who took her aboard a UFO. Once aboard, she met another friendly alien named Anuda, who put a "respiratory-type of device" over her face. The aliens appeared human except that their lips were very thin.

While on board, they offered her a sweet drink which she refused because of her diabetes. Another alien named Atari told her that she was now cured. Back at home, her doctor administered a blood test and pronounced the girl free of the disease. An interesting fact is that the girl was given some "mysterious stones" which she was told to place in a glass of water. She was instructed to drink the water "to assist in the cure."

Today the girl no longer suffers from diabetes.[15]

---

14. Morrow, Jan./Feb. 1993, pp. 15-17.

Humanity has struggled against a huge number of diseases. Fortunately, medical progress has allowed some diseases to be totally eradicated. Polio was one dreaded disease that caused near panic in the areas where it broke out. Parents kept their children home from school and streets became largely deserted. They knew the horrors of the disease firsthand. Polio is an infectious viral disease that affects the central nervous system.

Although not fatal, the disease causes muscle weakness and eventually, paralysis. There are several treatments, but there are no cures. Widespread inoculations have largely eradicated the disease. However, because inoculations are a relatively recent discovery, many people are still suffering from the ravages of the disease.

The following two cases both involve people who were cured of this disease by extraterrestrials.

CASE #083. Contactee John Adams of Toronto, Canada was stricken with polio of the spine when he was just an infant. He spent three years in the Shriner's Crippled Children's Hospital in Toronto, Canada. Doctors expected that he would die. However, they attempted treatment anyway. He was given gamma gobulin, and to the doctors' surprise, the treatment worked. John immediately began a slow recovery.

He had always thought that the human doctors were responsible for his recovery. It wasn't until many years later that John even realized he had been having repeated encounters with extraterrestrials. Under hypnosis, he was able to recall his memories of ET contact. It was then that he discovered who was really responsible for his remarkable recovery.

Under hypnosis, John recalled "scenes of 'people' around me talking ('He has only six months to live') and exposing a part of my body, turning to me and asking, 'Do you want us to take out the black vein or the red one?'"

John believes that this is a memory of being cured of his condition by aliens. As he says, "These I am discerning were contacts with other life forms." John now remembers over seven separate contacts, and says that the visitations continue even today.[16]

---

15. Morrow, Jan./Feb. 1993, p. 17.
16. Adams, Aug. 1992, pp. 11-13.

CASE #086. Contactee Richard Rylka has appeared elsewhere in this book because of the numerous cures he has received at the hands of extraterrestrials.

As a child, Rylka contracted polio in a public swimming pool in Newark, New Jersey. He was treated by his family with massage therapy. However, at night, he was treated by the aliens who would come into his room and hold a "flat instrument" over his body. They would manipulate the instrument which would flash brightly. After the treatments, Rylka discovered that his paralysis and weakness disappeared. Rylka is convinced that the aliens cured him of his polio.[17]

One condition from which millions of people suffer is infertility. Although not physically painful, this condition can cause untold mental anguish. Being denied the right to reproduce can be a devastating experience. Many people go through great lengths, including experimental and sometimes painful medical procedures, all for the chance of having their own child.

The aliens have shown an extraordinary interest in the human reproductive system. One of the central features of an extraterrestrial medical examination may be the removal of reproductive material. Some female abductees report mysterious pregnancies which vanish after a UFO abduction. Under hypnosis, these women often remember a medical procedure in which a fetus was removed from their bodies. In later abductions, they are shown these children.

There are three cases on record in which people were cured of their infertility after an encounter with a UFO. The first case, investigated by ufologist Brad Steiger, involves a woman. The second, involving a man, is one of the best known British abductions on record. The healing aspect, however, is virtually unknown. The third case was investigated by Barbara Lamb and is truly one of the strangest healings on record.

CASE #081. One day, Doriel X. of Chicago, Illinois, saw a UFO (date not shown). A short time later, two beings mysteriously appeared in front of her. The two beings said their names were Leita and Gamal, and said that they wanted to "incarnate" through Doriel. They told Doriel that she was "one of them" and that she would be perfect for bearing a child.

---

17. Randazzo, 1993, Chapter 11.

Unfortunately, Doriel had already been diagnosed with infertility. Her doctor had told her in no uncertain terms that she would never be pregnant.

Doriel was told by the aliens, however, that she would become pregnant in six months. Six months later, the aliens' prediction came true. Doriel was inexplicably pregnant. Nine months later she had a baby girl by natural childbirth. Because of the unusual conditions involving her baby's conception, she gave her the middle name of Leita.

Doriel has no choice but to believe that the aliens cured her of her infertility. She has the child to prove it. And as she says in her own words, "Physiologically speaking, I had been told that I would be unable ever to become pregnant."[18]

CASE #047. On November 28, 1980, police officer Alan Godfrey of the West Yorkshire Metropolitan Police Force in England was assigned to handle a peculiar problem. The station had received several calls about a herd of missing cows that had been seen in various locations.

Godfrey had been twice commended for investigative work involving murders, so this assignment seemed trivial. On the way to find the cows, he encountered a "large object blocking the road." At first he thought it was a double-decker bus, but upon closer inspection he saw it was a metallic saucer-shaped object with colored lights hovering five feet above the road.

He tried to radio the station, but his car's radio was dead. Godfrey had no police training for such an experience, but he quickly realized he was seeing a UFO as portrayed in science fiction movies. As he sat there, sketching the object, Godfrey experienced a "jump in time." He found himself further ahead on the road, missing twenty minutes of time.

He rushed to the station and returned with several officers to the scene. Although it had been pouring rain all night, the area where the UFO had hovered was completely dry.

They found the cows in a fenced area in which they appeared to have been deposited. There were no tracks in the area and the gate was locked at nights. As Godfrey says, "The only way they could have gotten there is if something had gone plunk and dropped them there."

---

18. Steiger, 1988, p. 152.

At first, Godfrey was ridiculed by his fellow officers, but when other officers in neighboring counties reported similar experiences, the ridicule seemed to stop. Godfrey then received ridicule from another source, his superiors. They advised him to simply resign. When he refused, his police car was taken away and he was given a bicycle.

Eventually, he underwent hypnosis and recalled some of his experience aboard the UFO. Godfrey's wife, upon hearing his experience, made note of a peculiar event.

It seems that a year earlier, Godfrey was called to a local disturbance. He ended up being severely beaten by three individuals. During the beating, Godfrey was kicked in the testicles, causing him to lose one testicle. The end result was that he was unable to have a normal sex life and was no longer fertile. As Godfrey says, "I was also told I would definitely be sterile for the rest of my life. I would not be able to father children."

One night, a few weeks after the accident, Godfrey's wife was awakened by a strange sound over the house. She tried to wake up Godfrey, but inexplicably, he would not wake up. After a few minutes, the sound left. She never knew how to categorize the experience until her husband mentioned his UFO encounter. Only then did she realize what probably happened. She believes a UFO was responsible for the sound.

But it was the next morning that they realized something else unusual had occurred. Godfrey seemed to be completely cured. As his wife says, "The next morning, Alan and I made love for the first time since he was beaten up. And then I became pregnant."

Alan, however, became angry and assumed his wife was seeing another man. After all, he was sterile and his wife was pregnant. What other way could this happen? But just in case, Alan went to see the doctor. The doctor was stunned and said that Godfrey was miraculously and impossibly cured. As the doctor allegedly said, "I don't know how it happened. Your condition has been completely reversed. The test shows you're fertile, back to normal."[19]

CASE #028. This case, investigated by Barbara Lamb, occurred to an anonymous female in Colorado in 1972. As Barbara Lamb reports, "There's another woman who is not a regular client, but I've had long conversations with her. She had all kinds of reproductive problems.

---

19. *Visitors From the Unknown*. (television program) Writer Michael Grais, producer Sharon Gayle, director Penelope Spheeris.

She was not able to conceive a baby which she and her husband really wanted to do. They had all the testing, and she just wasn't able to get pregnant. They did that for a couple of years. Then he woke up one night to find that there was this large being having sex with his wife next to him in bed, right next to him in the same bed. He started to get up and they sort of zonked him out. And he couldn't move, but he still had his eyes open and he could see what was going on. There were two beings in the room and they were big, tall. And one of them was definitely on top of his wife having sex with her. And she looked like she was deeply asleep but also having a tremendous amount of pleasure. Now both of them forget that for probably about nineteen years. And then the husband who had witnessed this started to remember. He did a regression about this and got the details of it. And then, of course, he told his wife, and she began to remember some of it too, and remember other encounters.

"To make a long story short, at a certain point, the extraterrestrials said to her in one of her experiences, 'Now you have given us two sons through this mating procedure. And now we will allow you to have two children.' So they did whatever they did, and cleared up her reproductive difficulties. And she was then able to conceive and bear two children, whom I met. They were teen-agers when I met them, and they are just fine. They're very human and normal."

Tumors are yet another condition that afflicts millions of people. Even benign tumors can be life-threatening. The following case documents how one man's tumor disappeared after a close encounter with an invisible entity evidently from a UFO.

CASE #082. An anonymous man from East Kilbride, England, reports that his house is haunted and that he had been abducted by grey-type aliens who claimed to be from the planet Venus, though on a higher vibration of existence. During one occasion, the witness was alone in his home when an unseen entity grabbed his arms and pulled them up over his head. He was unable to move and lost consciousness. When he awoke, he felt a "terrific warm glow in the chest region." He was then overcome by a feeling of peacefulness.

The anonymous man had been previously diagnosed with a tumor in the chest cavity. Some time later, the man went to the doctor and discovered that his tumor had "mysteriously disappeared." He attributes the tumor's disappearance to the visitation at his home.[20]

Another healing case involving tumors occurred to famed UFO abductee, Debbie Jordan (formally Debbie Tomey), who was featured as Kathie Davis in Budd Hopkins' book, *Intruders*.

CASE #075. In September 1993, Debbie Jordan of Kokomo, Indiana had been diagnosed with "female problems" including a large tumor. The doctors recommended surgery. Sometime after the diagnosis and before the surgery, Debbie was with three other people watching TV, when they saw a "big, bright, multicolored light" in the night sky. Grabbing the video camera, they watched the light for over ten minutes, capturing it on video tape.

When they returned inside, they went over what happened. It was then that, as Debbie says, "we realized we had missing time."

Debbie knew from past experience that missing time meant a probable close encounter. Later, she went to the hospital for her scheduled surgery. As she says, "I went in to have my surgery—they went ahead and did it because other things were wrong. But they had seen that tumor on the ultrasound, and then when they did the surgery, the tumor was gone."

Debbie asked the doctor about the tumor which had not only appeared on the ultrasound, but had been physically felt by her doctor. The surgeon replied, "There wasn't any tumor."

Debbie said, "What was it that he [Debbie's former doctor] saw?"

The surgeon replied, "I don't know, but there wasn't any tumor. Who knows what it could have been, but it's gone now."[21]

Another cure of tumors comes from California UFO investigator, William Hamilton.

CASE #078. This case involves a lady named Morgana Van Klausen from southern California. She has had several close encounters throughout the late 1980s and early 1990s. On December 4, 1994, Morgana went to the doctor and discovered that she had an unwelcome cyst. As the chief investigator of the case, Bill Hamilton says, "An X-ray found a peculiarity in the axillary segment of the right breast. This was followed by an ultrasonography over the same area. A small noncystic mass measuring 1.6 x .6 cm was found in the problem area."

---

20. Robinson, May/June 1993, p. 18.
21. *UFOs Tonite with Don Ecker*. Jan. 22, 1994.

Surgery was scheduled for December 14 so that Morgana could have time to recover from the slight fever she had developed. However, a day before the surgery, on December 13, Morgana was driving with her son when they saw a "white triangular-shaped craft" hovering nearby. They watched it for a moment until it suddenly darted away. Her son became very excited and said, 'Mom, look! Wow, this is a good sign—they are protecting us. We are protected. You wait and see, you'll have no more problems!'

That evening, Morgana was overcome by intense pain in the area of the noncystic mass. She tried to get up out of bed, but was unable to move. At that point, she lost consciousness. She woke up the next morning and went straight to the doctor to have her surgery. That's when both she and the doctor got a huge surprise.

The doctor inserted a needle into the mass and took another X-ray. To everyone's surprise, the new X-ray showed that there was no mass. The old X-ray was retrieved which clearly showed the mass. Another ultrasound was ordered, which confirmed that the mass had disappeared. As Bill Hamilton says, "Both the radiologist and the surgeon confirmed that the previous diagnostic had revealed a solid mass and that solid masses just don't disappear."

Hamilton has obtained the medical records which confirm the disappearance of the mass. He speculates that the aliens may have been removing an implant, however, as he says, "The explanation is only speculative." Whatever the case, Hamilton says that, "Morgana reported that she felt an unusual sense of well-being after the mass was gone."[22]

The following case also involves a cure of a tumorous cyst. I was able to interview the witness firsthand.

CASE #071. Alicia Hansen has had numerous contacts with extraterrestrials. She has undergone hypnotic regression to recall her early encounters, but now experiences full conscious contact. Back in 1991, in Atlanta, Georgia, however, she had an experience which she didn't fully understand until years later. It involved the disappearance of a cyst. As Alicia says, "It was really weird because at the time, I was pregnant. And I took a pregnancy test and it said positive. I didn't want to be pregnant and I was really upset. I went and I told my husband because we didn't have any money. I told him and I was

---

22. Hamilton, Aug. 1995, pp. 14-16.

crying. I said, 'Oh, my God! This is horrible!' And bla-bla-bla. And anyway, I got this cyst. So I was really worried about the cyst and being pregnant and all that. He's like, 'Don't worry, we'll work it out.' I was worried about getting health insurance and all that. I had already been diagnosed with the cyst by the doctors. I was having a lot of problems with it too. They did an MRI, where they take a picture and look at the cyst. So they saw the cyst and I was going to go back and have surgery on it.

"So time went on. I don't know what happened, but I woke up feeling really uncomfortable. Then I went to go have another pregnancy test at the hospital and they said, 'You're not pregnant.' I said, 'I already had a test and I was pregnant.' They said, 'We're really sorry, you're not pregnant.' And I thought that was really weird. And I was gaining weight and everything. I *knew* I was pregnant. And I took a test and it said yes, a big T. So I thought that was strange. I thought, 'Okay, I have just somehow spontaneously aborted or something.' I wasn't sure what happened.

"So I went back to the hospital to get another MRI done and make sure everything was okay. And the nurse said, 'Well, this is so strange,'—not the nurse but the person who reads the picture. They said, 'This is really weird because if we didn't know any better, we'd say that you never had a cyst. But what's going on here is there's fluid in your fallopian tubes. And the fluid, right where the cyst was, this fluid is not present when it ruptures. This is a fluid that is usually present after surgery. And that is when the fluid is there.'

'So it completely looked as if I had had surgery done, and I didn't. There was a pregnancy that had been taken, but I also had a cyst that was removed too."

Although Alicia doesn't consciously recall the surgery, the indications are pretty strong. Under hypnotic regression, she has recalled this surgery, and many others that she has had. She also received another cure of body injuries after a terrible auto accident. After having so many surgeries, she has become quite familiar with the medical technology of extraterrestrials, which far surpasses our own. As she says, "They use a lot of lights. They use a lot of lasers. And they use sounds too."

Alicia has had no further problems with cysts.[23]

---

23. Personal files.

Richard Rylka is unique among contactees in that he has received more extraterrestrial cures than any other person. He has been cured of injuries, illnesses and diseases. In the following case, he was cured of yet another condition which has caused him considerable pain.

CASE #058. On July 5, 1989, Rylka received a cure for a life-threatening illness.

In the mid-1980s, Rylka developed a tumor on the right side of his forehead. He was told by doctors that the tumor was related to a leukemia-type condition from his childhood. He had already had several benign tumors surgically removed from his body, yet the tumors kept coming back. Rylka's alien friends advised him not to have more surgery. Unfortunately, tumors kept reappearing, so Rylka was forced to have another operation.

Then, during his stay at a convention in Atlantic City, New Jersey, Rylka's alien friends, Koran and Nepos appeared and removed one particularly bothersome tumor, leaving no traces.

That seemed to solve the problem until years later, when the tumor appeared on Rylka's forehead. The tumor quickly became inches wide and showed no signs of healing. Friends and family began to express concern that it would affect his health.

By 1989, Rylka was also becoming concerned. Then, on July 5 of that year, Rylka received a house-visit by one of his ET friends. The ET told him that they were aware of his tumor and had been searching for a method to remove it.

The aliens took him aboard a UFO and told him that they were going to test a new healing device on him. They placed a small hand-held instrument that emitted a beam of light. Rylka was placed in an unconscious state and reports that he felt no physical sensation. When he awoke, the aliens told Rylka that his tumor was gone. He was instantly "beamed" back into his bedroom.

Once in his bedroom, Rylka dashed to the mirror and saw that the aliens were right; his tumor had vanished. Nearly one hundred people had witnessed Rylka's tumor, so when it disappeared, it amazed a great number of people, all of whom verified the cure. Rylka states that many people seemed surprised, making comments like, "My God, they're not all bad, are they?"

Today, Rylka continues to maintain contact with his extraterrestrial friends.[24]

---

24. Randazzo, 1993, Chapter 11.

Still another condition that affects many people is paralysis. Being able to walk is something many of us take for granted. For other people, this simple act is impossible. In the following case, a wheelchair bound man was taken inside a UFO where he experienced a remarkable healing.

CASE #051. One evening in 1986, Richard T. of La Jolla, California, was alone on the beach sitting in his ever-present wheelchair when a "hundred-foot-long, torpedo-shaped UFO" appeared above him. The next thing he knew, Richard found himself inside the object, surrounded by small humanoids with huge heads and large, slanted eyes. To his surprise, Richard felt no fear, only very peaceful.

His next memory is of waking up in the back of his van, with his wheelchair folded beside him. Richard didn't know it, but he had been cured. Brad Steiger, who investigated the report, writes, "Over the next few weeks, his condition began to reverse itself—until he was finally able to walk again with the help of a cane."[25]

In the following case, the witness was cured of an illness that he didn't even know he had! According to the aliens, the illness was life-threatening.

CASE #077. "David," an American electrical engineer, was abducted from his bedroom on September 12, 1994. He had no memory of the abduction at all. His only clue was a large triangular bruise on his abdomen. Doctors were baffled. David eventually elected to undergo regressive hypnosis during which he recalled being taken out of his room by grey-type aliens in robes up into a triangular-shaped craft. He was put onto a table and examined. David had the sense that they had an important job to do. As he says, "There seems to be an urgency right now for what they're doing.... There is something very wrong with me...they're worried...very worried. Like they're showing me images of myself...dead...with blood coming out of my mouth. Some sort of urgency!"

During this time, David felt the aliens performing complex surgery on his abdomen. As he says, "I feel like something is being pulled...uh...open. Like my abdominal area is being stretched. It's very hot like being under a very intense lamp. They don't want me to see anything. I feel a dull pain...I'm paralyzed. I'm motionless. It's

25. Steiger and Steiger, 1994, pp. 40-41.

like I'm stuck to a large magnet...and you couldn't move a finger if you tried...There's a hand on my forehead, covering my eyes...holding me. I feel almost as if something is stretched over it, there's no pain now, no pain whatsoever. It's just a weird feeling. Kind of like the skin is being stretched. There's no pain...whatsoever...like it's like pain is being totally blocked."

David had experienced UFO encounters before and was always left feeling anxious. This encounter gave him hope. As UFO investigator C. Leigh Culver, who investigated the case, writes, "Before this episode, David had experienced some intestinal difficulties and he felt as though the beings had helped him with this in some way."[26]

As is usually the case, there are reports of a number of other healings of serious diseases. The problem is that many of these reports do not contain detailed information. The following reports may be incomplete, but again they show how widespread UFO healings actually are.

CASE #105. The reported abductions and healings of Linda X. of California have been recounted elsewhere in this book. Linda witnessed the healing of many other people. One of the healings she saw involved a man who was cured of his tumor.

She saw the man with the tumor lying on a table. The aliens held a small, circular item over the man's tumor, which quickly began to disappear. As Linda says, "They're pulsing it to shrink the tumor, and I can almost see the tumor shrinking as they're working on it." After the tumor was gone, the man got off the table, and another patient came to the table to be examined for health problems.[27]

CASE #091. Contactees Bert and Denise Twiggs of Hubbard, Oregon, have received many cures at the hands of their alien friends from Andromeda. In fact, each member of their family has been cured at one time or another.

One cure which was mentioned only briefly is nevertheless quite astonishing. Denise Twiggs explains how she suffered from colitis. The case was so severe, it actually caused a section of her intestines to bleed and become badly inflamed.

Denise reports that at this point, she was taken aboard the ship into presumably the alien examination room, where as she says,

26. Culver, Vol. 2, No. 3, pp. 3-7, 29-30.
27. Fiore, 1989, p. 105.

"...they removed a section of my intestines which had been ulcerated by colitis."[28]

CASE #031. In October 1973, in a small town in the mid-western United States, Pat Roach and three of her seven children experienced a simultaneous UFO abduction. All of them except the youngest experienced a period of missing time and only remembered the experience under hypnosis.

This case was investigated by several well-known UFO researchers, including James Harder Ph.D., Captain Kevin Randle, and Jim and Coral Lorenzen. They performed hypnosis on the witnesses, and a typical UFO abduction scenario was revealed.

The youngest, Debbie "Dottie" Roach, however, already had full conscious memory of the event. Right after the event occurred, Pat assumed there had been a prowler. Debbie said simply, "They were spacemen...they didn't make me forget. They told me not to tell anyone except those in my family."

Debbie then gave a detailed description of the aliens, the ship and what happened when they were all taken onboard. According to Debbie, they were all told to get on a machine. Debbie, who had been very ill all her life, was told by the aliens that she would be healed. As Debbie says, "The one that stood in the corner asked my name. And he said that I wouldn't be sick anymore."

As the report says, "Pat had said that Debbie had been very sick before the aliens arrived and they had done something to or for her. The sickness was gone after the aliens left."[29]

As we have seen, people have been cured of all kinds of diseases by extraterrestrials. Many of these diseases are considered to be chronic and/or incurable. Obviously, it is our own limitations that cause a disease to be labeled incurable. It may be that there is no such thing as an incurable disease.

One disease that is conspicuously absent from the UFO cures is AIDS. As far as I know, there is only one case in which somebody was cured of the disease. Unfortunately, the cure is not verified. The case was researched by southern California hypnotherapist Yvonne Smith. At the 1992 UFO Abduction Conference at M. I. T., Smith

---

28. Twiggs, 1995, p. 188.
29. Randle, 1988, pp. 17-24.

spoke briefly about the case saying, "I have an HIV-positive abductee who now tests negative."[30]

I contacted Smith regarding the case and she graciously supplied further details. The gentleman had tested positive for the disease and then, after experiencing a few abductions, tested negative. Smith asked for medical records, but the gentleman moved to the eastern coast of the United States and she lost contact with him. The last time she saw him, however, he appeared to be in good health.

There is also at least one case in which information was given about the disease. Not surprisingly, the receiver of the information was contactee Richard Rylka. His contacts told him only that they were experimenting with introducing hydrogen peroxide into the body's system. Rylka has no idea how this is done or what the effects might be.

---

30. Bryan, 1995, p. 20.

KESARAG

# Cancer Cures

Probably the most dreaded of all diseases is cancer. Unlike most diseases, it is not limited to one organ or organ system. It can strike any part of the body at any time. It is also one of the most unprejudiced diseases, striking people of different ages, sexes and backgrounds. It attacks without rhyme or reason, creating havoc inside the human body.

Although there are some known causes of cancer, many other cases arise spontaneously and seem to have no particular cause. Of course, cancer can also spontaneously remiss. However, remission without treatment is extremely rare.

Cancer often kills people quickly and with few warning signs. The chances of recovery are usually not very good, unless the disease is caught before it metastasizes and sends cancerous cells throughout the blood system and entire body. Unfortunately, metastas can occur before any obvious signs of the disease make themselves apparent.

Cancer is an insidious disease because, unlike other diseases, it turns the body against itself. It can be basically defined as the abnormal and uncontrolled division of cells which invade and destroy surrounding tissues.

Cures for cancer range from the absurd to the tortuous, and none of them are really cures in the true sense of the word. In actuality, they are only treatments. And these treatments are not necessarily effective against the disease; there are many cases which simply do not respond. Normal cancer treatments include chemotherapy radiation, surgery, and radical changes in diet and smoking habits.

Because cancer is such a widespread and treacherous disease, it would seem logical that the aliens would have a strong interest in being able to cure it.

This turns out to be true. In fact, cancer cures are one of the most common types of UFO cures. And as we shall see, the aliens have cured cancers of virtually all types.

In the following case, the aliens were able to help cure someone of a cancerous skin condition.

CASE #046. On November 19, 1980, art teacher Michael X. and his wife Mary of Longmont, Colorado, were driving late at night when a bright blue beam of light shone down on their car from a hovering UFO. They felt the car being lifted off the road, at which point they suffered a period of missing time.

Under hypnosis, they remembered being taken inside a craft piloted by short, gray-skinned beings dressed in shiny jumpsuits. Both were given the standard physical examination. Afterwards, they were released back into the car and found themselves moving along at 55 miles per hour.

Mary was two months pregnant at the time but suffered no apparent physiological effects. Michael, however, experienced an apparent healing of his mild case of skin cancer. As the report says, "Michael had several large suspected cancerous melanomas on his legs before the incident; afterwards they were greatly reduced in size and coloration."[1]

Skin cancer kills many people. Fortunately, because it is highly visible, it is easy to detect early in order to begin effective treatment. Other forms of cancer, however, are not very apparent until they have spread throughout the body. In the following case, investigated by long-time UFO researcher Timothy Green Beckley, a young lady suffered from cancer that slowly metastasized through her entire body. At the point where she probably would have died, the lady was cured in an amazing way.

CASE #036. In May of 1974, Helen X. of Arizona was diagnosed with cancer of her hip-bone. Doctors quickly had the cancer surgically removed. A short time later, a cancerous growth was discovered on her pancreas. The doctors again operated on her and removed most of the cancerous material. Unfortunately, there were unforeseen complications. As Helen says, "They were not able to remove all of it, so they just closed me up."

---

1. Hall, 1988, pp. 311-312.

A year later, the cancer had spread to Helen's bowels. Helen went through chemotherapy and lost 50 pounds. The treatment did not stop the spread of the cancer, and doctors told her in no uncertain terms that she would die soon and that she should see her family before it was too late.

Helen was so weakened by the disease that she was barely able to walk. One evening, she was too tired to even eat and just went to bed. At 1:00 A.M., she woke up to hear her name being called. She had an overwhelming impulse to get into her car and drive to a certain location about seven miles out of town.

She was unable to resist the impulse and quickly drove to the spot. To her surprise, she saw a large, white UFO hovering in the sky. The object floated to the ground, and two small creatures in tight-fitting metallic suits exited the craft.

Helen was quickly taken on board and was told not to be afraid, as the aliens only wanted to help her. Although she was frightened, Helen felt that the aliens were friendly. She was laid down on a table and examined with a small reddish instrument shaped like "an upside-down mushroom." The aliens passed it over her body several times and then told Helen that she had cancer, as if she didn't already know!

The aliens, however, were very specific. They told Helen that she had cancer in her left breast, liver, right kidney, pancreas and spleen.

Then began the long and sometimes painful cure. First, an instrument shaped like a "metallic tray with a handle on either side of it" was passed over her body at least ten times. Each time the instrument passed over a cancerous area, Helen felt intense heat and excruciating pain.

Helen was able to withstand the tortuous operation only because the Earth doctors had already done much worse. After the traylike instrument, the aliens pulled out a foot-long tube filled with "purplish-colored liquid." They took this tube and injected the fluid into Helen's abdomen. A short time later, they injected her with another tube and took out several ounces of dark-colored blood. Helen was then given several more injections on her abdomen, sides and back. Then more strange instruments were passed over her entire body.

Afterwards, Helen was told she could get dressed. The aliens showed her a map and explained that they were from a solar system beyond Orion. They told her to take no medication, and that her cancer was completely cured.

Helen drove quickly home and went straight to bed. The next morning, she was still very weak and appeared to be suffering as badly as ever. She told her son of her experience. He was skeptical until they drove to the location and found a large depression in the earth where Helen remembered the UFO had landed.

That afternoon, Helen began vomiting foul-smelling "black-stuff." She was rushed to the hospital, where doctors gathered Helen's family and told them she was about to die.

For two days, Helen was violently ill, and lapsed in and out of consciousness. Doctors offered numerous medicines, but Helen, following the orders of the aliens, refused all medications.

By the third day, Helen had recovered. Nearly all traces of the cancer were gone. As she says, "It was just like I had never been sick."

Two weeks later, Helen's skin color returned to normal and the doctors had to admit that she was miraculously cured. Her cancer and subsequent cure are verified by doctors and Helen's medical records.

Today, Helen is perfectly healthy and manages her own business. She also says that, when the time is right, she will go public with her experiences.[2]

A similar case comes from UFO researcher Antonio Huneeus, contributing editor to *UFO Universe* and columnist for *Fate*. About this particular case, Huneeus says, "Perhaps one of the most interesting UFO cases not only in Peru, but in my files from anywhere in the world, involves a truck driver who was allegedly cured of a malignant stomach cancer following a close encounter with a UFO that emitted a powerful beam of light."

The case was initially investigated by ufologist Anton Ponce de Leon of Peru. Unfortunately, there is no date provided for the incident.

CASE #085. Truck driver Mr. X. of Peru was driving in the evening when he suffered two consecutive flat tires outside the city of Santa Rosa. He called in for help.

At the time of the drive, Mr. X. was suffering the intense pain of stomach cancer. His doctors had been urging him to have an operation, but Mr. X. remained hesitant. However, while waiting in pain for help to arrive, he decided he would go ahead with the operation.

---

2. Johnson, July 1988, pp. 25-26.

Later on, the sky was suddenly lit up by a bright object. Mr. X. exited the truck and looked up in the sky. To his surprise, he was engulfed in a bright beam of light. The driver then experienced a period of missing time.

Shortly later, the UFO left and help arrived. When Mr. X. returned home, he discovered that he was no longer suffering any stomach pains. He decided to see the doctors once again. According to Ponce de Leon, "...he still had doubts and asked for a new analysis and X rays. To the doctor's surprise—he still doesn't know what happened—the stomach cancer had disappeared and, so, naturally, he was bothered by the entire situation and was unable to give his patient any definite answers. Mr. X. remained with his secret and the doctor continued to be upset because he never did find out what really happened. I investigated this case first-hand and I know it's a very important one!"[3]

Yet another cure of cancer caused by a beam of light comes from French UFO investigator Gilles Garreau. It involves a lady who was cured of throat cancer.

CASE #032. In 1973, 49-year-old factory worker Denise B. of Marseille, France, was suffering from a worsening case of throat cancer. She received regular treatments of chemotherapy. Unfortunately, the treatments did little to halt the disease, so she slowly discontinued the treatments.

Meanwhile, the cancer got worse. Denise had become partially paralyzed and had no reflex actions left.

By 1974, she had stopped all treatments. She still visited the hospital every month. On August 15, 1974, her doctors stated that the cancer was worse and that she must return to the hospital on September 15.

On the evening of September 14, Denise B. was taking a walk with her dog. The dog disappeared into some bushes and Denise followed, dragging her feet because she could barely walk.

Suddenly, Denise was overcome by a feeling of "having her insides all mixed up." Shortly later, she felt strangely tired, and her flashlight stopped working. She then felt something grip her neck. She noticed that her whole body was engulfed in a beam of light com-

---

3. Huneeus, Spring 1994, p. 52.

ing from above. She was unable to move and felt the beam lift her into the air.

Her next memory is of seeing another light near the ground. Then the lights disappeared, and Denise returned to the road. She noticed that she could walk fine and her legs no longer hurt. She recovered her dog and returned home, thoroughly frightened.

The next day, she went to the hospital. According to the report on the case, "The next day in the hospital, the medical assistant examined her: no more cancer! This was confirmed by another doctor and she was told she could go home."

Unfortunately, the doctors did not give Denise her medical records. She told one mystified doctor that it must have been the goats' milk she drank that cured her.

Today, Denise B., at nearly 70, is healthy and totally free of cancer. Although she is not religious, she does believe her experience was unusual. As she says, "If only this miracle could lead to a cure for cancer."[4]

There are many treatments for cancer. However, as we have seen, these treatments do not always work. Doctors are then given the heart-wrenching task of telling a patient that his/her condition is terminal.

But as we have also seen, aliens are quite capable when it comes to curing even terminal cancer. In the following case, the patient was diagnosed with cancer and was told she had only three months to live. Unknown to her doctor, however, she had been having onboard UFO encounters all her life, including one involving a healing.

CASE #067. Licia Davidson of Los Angeles, California, has been having UFO contacts for as long as she can remember. In 1989, Licia was diagnosed with terminal cancer. By the time the disease was diagnosed, it had already metastasized to her colon, making it inoperable. She was given three months to live.

She then experienced an abduction, during which she was placed on a table and given an extensive operation to cure her cancer. As she says, "I was abducted. They told me I had cancer. They said, 'Relax.' And they did a cure. It was excruciating."

Licia was returned to her Los Angeles home. Upon returning to her doctor, it was discovered that all traces of the cancer were gone.

---

4. Mesnard, Jan. 1994, pp. 10-11.

Licia has recovered her medical records, and states that her cure has been verified by a major medical university. She also states that she fears the United States Government (who has harassed her extensively) more than she does the aliens.[5]

Generally speaking, being cured of an illness by a UFO encounter is extremely rare. Most people are able to visit only human doctors. And even those who do receive a healing by a UFO know that they cannot depend upon the aliens to keep them healthy.

There are a few cases on record, however, in which people have received multiple healings. The following case involves a lady who was cured of a yeast infection and a liver infection as well as two forms of cancer.

CASE #097. Abductee Linda X. of California did not know that she had cancer when she was abducted out of her bedroom into a UFO, placed on a table and operated on by aliens. They told her she had cancer and that they would cure it for her.

She watched as the aliens opened up her stomach and removed the cancerous material. As she says, "I don't understand what they're doing. There's no blood. They just open me up. It looks real dark inside. They're suctioning something up...kind of cleaning me, inside."

The cancerous material was quickly removed from Linda's abdominal area. The stomach was then closed up, leaving no scar. Afterwards, strange instruments that looked like "little round pieces of glass" with lights on them were placed at various points upon Linda's body. Another instrument was then passed over her body. When the instruments came to the cancerous area, Linda reported feeling intense pain.

She was then quickly deposited back into her bedroom.

On another occasion, in the late 1980s, Linda was given another cure, this time for an apparent cancerous tumor in her left breast. She was again put on a "long table, like in a doctor's office."

The aliens then took a "metal tube with a light on the end" and held it over Linda's left breast. Some form of energy was emitted by the object which pulsated and glowed. The tube was then removed and Linda was told to get off the table to make room for another patient.

---

5. Davidson, Licia. 1994 UFO Expo West, Los Angeles, CA.

Again, Linda has seen numerous other people being cured aboard the UFO, including her own family. In fact, she remembers that her sister, Sherry, was operated upon for an unidentified illness. Afterwards Sherry jokingly asked the aliens if they could also do breast enlargements!

Obviously, both Linda and Sherry felt very comfortable and welcome aboard the alien craft.[6]

Cancer is a very dangerous disease, yet aliens seem to be able to cure it relatively easily. Cancer can, of course, go into spontaneous remission, making it difficult to pinpoint the actual cause of a cure. The following case involves a cancer cure that may or may not be attributable to a UFO experience.

CASE #052. In the mid-1980s, Edward Carlos of Pennsylvania discovered a mark on his lower body that was diagnosed as cancerous. Carlos immediately had an operation and the cancer went into remission. However, Carlos attributes the remission to the energies he experiences through gardening and water-color painting, and to the "energies transmitted to him by the abduction process."[7]

It may seem that contactees are lucky to have alien friends that cure them of various illnesses and diseases. However, being a contactee is not all fun and games. Like many people who report UFO encounters, contactees are often accused of lying and are even shamelessly ridiculed. When contactees go public with their experiences, they can leave themselves wide open to this kind of attack.

Usually contactees are chosen in order to fulfill a certain task, and they are often expected to tell people of their experiences, despite ridicule or disbelief. Of course, contactees must maintain good health to complete their mission. The Twiggseses, with whom you should be familiar by now, are just a few of the many contactees who have gone public with their experiences. Among the many cures they have received, there is also one healing of cancer.

CASE #092. Elizabeth X. is one of the many Earth people that Bert and Denise Twiggs have seen aboard the Androme mothership. Elizabeth witnessed Bert being cured of cancer he didn't even know he had. Elizabeth clearly remembered that the aliens diagnosed Bert

---

6. Fiore, 1989, pp. 89-114.
7. Mack, 1994, p. 46.

with lung cancer. They promptly performed surgery, and removed the cancer from the lung in which it had appeared.[8]

Another cancer case comes from well known UFO researcher Wendelle C. Stevens, who is responsible for bringing a number of contactee cases to public attention.

CASE #084. An unidentified Hungarian woman told UFO researcher Wendelle C. Stevens that she was a contactee. Stevens reports that the woman told him that she had been taken aboard UFOs on four separate occasions. While on board, she claims to have met friendly aliens who cured her of cancer. She refused to go public with her experiences after being warned by Hungarian government officials to remain silent.[9]

The cancer cures are the most dramatic and unbelievable of all UFO cures. They seem to involve complex procedures with many instruments. Beams of light are often used, as they are on other illnesses. Sometimes open surgery is required. In other cases, pills are given.

The case which follows involves all these methods of healing. Evidently, the cancer was bad enough that it involved extensive attention. The case gripped the entire UFO community when it was first revealed by Brazilian journalist Joao Martins and UFO investigator Olavo T. Fontes, who was the official medical investigator of UFOs for the Brazilian government. It was also published in an article by UFO investigator Gordon Creighton in the popular journal, *Flying Saucer Review*. Today, the case has become one of the most well known and widely reported of all UFO healings.

CASE #011. On October 25, 1957, an anonymous family in Petropolis, Brazil was visited by a UFO which cured one member of the family of terminal stomach cancer.

According to the report, the family was very wealthy, and the father was a powerful and influential figure in society. The man's young daughter was suffering from stomach cancer and doctors expected that she would die. The daughter's disease progressed quickly. She was suffering intensely on the night the UFO visited the family.

---

8. Twiggs, 1995, p. 188.
9. Condon, Oct. 29-Nov. 4, 1992.

What follows is the first-hand testimony of Anazia Maria, who was the family's maid and an eyewitness to the event. It appeared originally in the book, *UFO Abduction At Mirassol*, published by Wendelle C. Stevens of the UFO Photo Archives.

It was the following letter which first brought the case to the attention of the UFO community.

"Dear Sr. Martins: I have read your articles and desire to compliment you on them. I believe in the existence of the flying disks because I personally witnessed the incident here related. I don't know if you will believe me, but I assure you that what I am going to tell you is the truth. I am poor but honest. I will not mention the whole names involved but I am sure you will understand.

"My name is Anazia Maria. I am 37 years old and actually live in Rio De Janeiro. I have worked in the home of Sr. X. since December 1957. You must excuse me for not mentioning his name for he is a wealthy and influential man of this city. The daughter of my employer suffered from cancer of the stomach. She suffered considerably, and I was employed to serve as a sort of governess, and to assist the Senorita Lais, the daughter with cancer.

"She had been given many and diverse treatments to control the cancer, but the doctors had said there was no hope. In August of 1957 my employer sent the whole family to a small farm near Petropolis, hoping his daughter would improve in the better climate. But the days passed with no improvement. Lais could no longer eat and her suffering was horrible and became worse every day. She was given constant injections of morphine to control the pain.

"I remember the night of 25 October very well. We expected her to die as the pains of Senorita Lais were terrible. The injections seemed to have no effect. Her father was singing a verse when suddenly a strong bright light came on to one side and shown directly on the house. We were all in the room of Senorita Lais whose window was situated exactly on the side of the house from which the light came. The only light inside was a small lamp at the head of the bed.

"'Look,' cried Julinho, a brother of the dying girl, as he ran to the window and saw a disc-shaped machine there in the light. It was not very big, and I am not learned enough to be able to say what its diameter might be or its height. I know that it was not very big. The upper part was enveloped in a luminescence of a yellowish-reddish color.

"Suddenly, a port automatically opened in the object and two small beings emerged. They came toward the house while a third one remained in the port of the disc. I noticed on the interior (of the disc) through the port I could see a green light dimly like can be seen in 'night clubs.'

"The 'humans' came into the house. They were short, about 1.20 meters tall, smaller than a younger son of my employer who was 10 years old. They had large heads on top of their shoulders, reddish ears, small slanted eyes (like Chinese,) but their skin was a vivid green!

"They had something on their hands that I thought were gloves. Their suits were white and seemed to be thick—the chest, the sides and the cuffs shined brightly. I do not know how to explain. They came up to the side of Senorita Lais, who moaned in pain, eyes wide open, and not understanding anything that was going on around her. Nobody moved or spoke in the terrible tension. I was there in the room with Sr. X. and his wife, Sr. Julinho and his wife, and Otavinho, the younger son of my employer.

"Those 'humans' looked at me silently and stood at the side of the bed and Lais placing the instruments that they carried on the top of the milky coverlet. They made a gesture towards my employer, and one of them placed his hand on the forehead of Sr. X. and 'discussed' the case of Senorita Lais with a bluish light that showed all of her interior. We saw all there was inside the belly of the girl. With another instrument that emitted a noise, 'he' pointed it in the direction of the stomach of Lais and we could see the cancer. This operation lasted about a half hour. Then Senorita Lais went to sleep and 'they' left. But before leaving the house, they communicated with Sr. X. telepathically that he would have to give her medicine during one month. Then 'they' gave him a sphere made of stainless-steel-like metal which contained some 30 small white pills. She was to be given one per day and she would be cured.

"Lais later returned to her doctor who verified that her cancer had been cured.

"I have left that house, but I promised to absolutely guard the secret of this case. Therefore I have told you, but I must guard the true identity.

"If you mention this case in your articles it is of no consequence because I have withheld the names of the people involved in this. But I assure you that everything I have said is real.

"Lais was condemned to die of cancer of the stomach and, in spite of this she was saved by an instrument that looked like a flashlight, that emitted rays that 'dissolved' the cancer, and she survived. They saved Senorita Lais and the same night returned to the flying machine and were gone."

The obvious sincerity of the letter and the fact that it closely matches up with other reports strongly indicate that this incident probably happened in the manner described. Despite the case's obvious fantastic element, the case is actually no more unique than any other. Nearly all the details have been reported in one case or another.[10]

As we have seen once again, the aliens really know their medicine. Where human doctors give up all hope for their patient's recovery, aliens intervene to cure illnesses in a matter of minutes. In some of the cases, the cancer was cured only when the witness was at the brink of death. In others, the patient didn't even know that he/she had cancer until told so by the aliens! Either way, it proves again that aliens are quite skillful physicians.

There are doubtless many other cases of cancer cures that remain unknown simply because the witnesses do not wish to come forward. As UFO researcher and psychologist Dr. Richard Boylan wrote in a letter, "I cannot release case names or identifying information. I can tell you in a general way that one woman I have dealt with had a healing of a vaginal yeast infection during an extraterrestrial encounter aboard a UFO, and subsequently, during another close encounter, apparently was cured of a cancerous growth on her uterus, which, when later examined by the surgeon, could find no trace of the previously diagnosed cancer."[11]

There is at least one case on record in which the witness, who was perfectly healthy, was simply told by the aliens of their ability to cure cancer. This happened to Kathleen Compardo of Pennsylvania in 1950. She was lying on her bed when two transparent figures appeared at her bedside. The figures were about three feet tall and told Kathleen that they were friendly aliens who wanted to help earthlings, but were afraid of their warlike ways. They told Kathleen that they were "far advanced in the fields of science and could cure cancer with a simple machine...."[12]

---

10. Buhler, Pereira and Pires, 1985a, pp. 139-141.
11. Personal files.
12. Holzer, 1976, p. 70.

In another case, contactee Cynthia Appleton of West Midlands, England, was given much knowledge of a technological nature by extraterrestrials. They gave her a report on the 'nature of cancer' and told her one effective method of treatment. This cure involved "shocks to the body organ most in use at the time, creating a sort of frequency change in the vibrational rate of the particles inside the atom."[13]

There are other cases on record in which the aliens have admitted that they know how to cure cancer. In the following case, aliens didn't actually implement a cure, but they told the abductee one method of curing cancer. Whether or not the cure actually works remains a matter of controversy. I have included the case because the cure given by the aliens was given to human cancer patients, and there appeared to be an improvement in their condition.

CASE #010. In the mid-1950s, "Sara Shaw" experienced several abductions from her home in Tujunga Canyon, California. After the encounters, she developed an obsessive interest in medicine and began going to medical school.

Years after the abduction, she underwent hypnosis to recover her last memories. She recalled that during one of the encounters in 1953, she was given directions on how to cure cancer. The aliens told her that cancer was basically a problem of "rotting." She was told that the cure was not for herself, but that in the future, she would meet a certain doctor whom she would recognize. She was instructed to give the cure to this doctor.

It is the information they told her that has caused many people to be skeptical of this cure. The aliens told Sara that cancer could be cured by swabbing vinegar directly on cancerous growths.

A few years later, Sara was very surprised and quite excited to recognize the doctor to whom she was supposed to give the cure. Coincidentally, her friend and co-abductee, Jan Whitley, was suffering from breast cancer at the time and it was her doctor whom Sara recognized as the right man. She was even more surprised when the doctor became very interested in her story and promised to try the cure on some of his patients who were looking for alternative methods of healing.

According to the doctor who performed the UFO-prescribed cure, he said that "...he had repeatedly applied vinegar to some skin

13. Randles, 1988, pp. 72-73.

cancers by using a swabbing technique and had indeed procured some success with this method." The doctor did indicate, however, that his patients were also undergoing numerous other treatments to cure the cancer, and that the improvement may have been attributable to another treatment.

Nevertheless, the results were intriguing. Further study by the investigators of this case revealed that according to folklore, vinegar was known to be a cure for many other ailments. Also curious is that some cancerous tumors may actually secrete acetic acid, the scientific name for vinegar. The doctor had originally told Sara that acetic acid couldn't possibly cure cancer as vinegar is already found around most cancerous tissues and that the cancers may actually secrete the substance. Sara astutely pointed out that the presence of vinegar around cancerous tumors could just as well mean that the body was producing the vinegar in an attempt to fight the cancer.

In an interview, the doctor also admitted that he had another patient who had the same revelation as Sara. This man also told the doctor that aliens had told him that vinegar is an effective cure against cancer. The doctor knew that cancers thrive in an alkaline medium, and since vinegar was acidic, it also seemed worth trying.

Thankfully, there has been some research in this area. The investigators of the case brought it to the attention of a physician who specializes in cancer immunology. He told them that vinegar alone has failed to cure cancer. However, he says that a chemical closely related to acetic acid but much more potent "is currently being used on skin cancer with significant results."

Hopefully, there will continue to be more research in this area.[14]

Obviously aliens are interested in our bodies and the welfare of our bodies. In fact, their interest in this area may be greater than many people realize.

---

14. Druffel and Rogo, 1988, pp. 57, 118-134.

KESARA

# Experiencers and Psychic Healing

Unexplained healings are now a recognized symptom of an encounter with a UFO. Another lesser known symptom of many UFO abductions is the fact that afterwards, the experiencers themselves become psychic healers.

This bizarre turn of events has turned up in so many cases, that there is little chance of it being a coincidence. There are a large number of people who have been abducted and then acquired either an interest in various forms of medicine or the psychic ability to cure people of various diseases.

One researcher who noticed this pattern is Gordon Creighton, who spent considerable time researching UFOs and psychic healers. He even underwent minor psychic surgery himself. Creighton observed healings being performed in which incisions were mysteriously made and healed, all by mysterious methods. As Creighton says, "The body is laid open paranormally, as if the operator were opening the magnetic body, unzipping it, so to speak. There is almost no blood. A few passes—still never touching the flesh—then he zips it up again."[1]

This sounds very similar to some of the UFO accounts of healing. Creighton thinks there is a connection. As he says, "It's an intriguing coincidence that in both Brazil and the Philippines, all of these healers began to develop their powers around 1947 and 1948—*after* the modern wave of sightings began."[2]

John Mack has also noticed this pattern. As he says, "Many abductees seem to gain healing powers themselves. Although abductees may continue to resent the abduction experiences and fear their reoccurrence, at the same time many in one way or another come to

---

1. Blum and Blum, 1974, p. 150.
2. Blum and Blum, 1974, p. 150.

feel that they are participating in a life-creating or life-changing process that has deep importance and value.... In addition, many abductees...appear to undergo profound personal growth and transformation. Each appears to come out of his or her experiences concerned about the fate of the earth and the continuation of human and other life-forms."[3]

Consider the following case in which an entire family was taught how to perform psychic healings. Linda X. of California has been cured several times by aliens. She saw other cures being performed, and she also saw her family aboard the UFO. But most importantly here, she says that the aliens are teaching her and her family how to perform psychic healings.

Linda remembers being taken to a room inside a UFO that contained over 60 people. There was a large crystal in the room that emitted a strange type of energy. The group was told by the aliens that they were being taught. Linda felt a strange electricity flowing through her body and out her hands. Then the aliens explained. As she says, "I'm being told I can learn to heal, myself. It's okay to use this power."[4]

Another less famous case is that of Michael Bershad, whose story is presented in Budd Hopkins' book, *Missing Time*, under the pseudonym, Steve Kilburn.[5] In Bershad's abduction, his back was opened up by the aliens who then proceeded to activate certain nerves as if they were testing his nervous system. Afterwards, they closed up his back, leaving no scars. However, Bershad has stated that ever since his abduction, he has had a "bad back."

Bershad does not feel that his aliens were friendly, and he has experienced subsequent contacts with them. Because of this, he has been exploring several techniques to keep them away and has had some recent successes in this area.

In 1993, Bershad spoke at the Los Angeles chapter of MUFON. At the meeting, one attendee asked if there were any positive effects caused by his experiences with the aliens. Bershad stated without hesitation that as a result of his abductions, he had become a psychic healer. He was even able to provide examples of this ability. As if it were almost normal, he stated that he had cured two ladies of cancer.

---

3. Mack, 1994, p. 398.
4. Fiore, 1989, p. 96.
5. Hopkins, 1981.

The cures were performed over the phone. It seems that many psychic healers are able to work from a distance.

Another interesting case is that of Joyce Bowles of Winchester, Hampshire, England. Bowles, a British Rail powder-room attendant was abducted on three separate occasions in 1976 and 1977. The case follows the standard patterns, including some paranormal effects. Not surprisingly, Bowles has a record of psychic experiences and is well-known in her neighborhood for her ability to cure animals.[6]

The area of Tblisi, Georgia, in Russia should be well known by now for the large number of cures that have occurred in the area. There are also a few cases where witnesses have become psychic healers. One lady became a healer after an entity entered her body, giving her the ability to see where a person's illness is manifesting, and then guiding her to administer a psychic cure.[7]

Another Tblisi case is that of Professor Bochereshoni, a mental health doctor. While working in the hospital, the doctor was visited by aliens who assisted him in diagnosing and curing the patients with mental illnesses. According to the report, the aliens also provided written formulas to cure a variety of illnesses.[8]

Several of John Mack's clients also report becoming involved in alternative forms of healing. Scott of Massachusetts is one of the abductees in Mack's book. According to Mack, "Scott himself, in addition to his increasing curiosity about the spiritual dimensions of the phenomenon, has begun to meet with an acupuncturist, and more recently, with a shamanic healer. He is also increasingly challenging the traditional model."

Scott himself says of some therapists, "I feel I could heal, I could help them more than they could help me...."[9]

Paul of New Hampshire is another typical case. He has experienced many onboard UFO encounters, and has become increasingly interested in healing. Paul was given volumes of ideas concerning "healing technologies" by the aliens. He has also become a healer himself. As Mack says, "Those who know him outside of the therapeutic setting, such as Pam, Julia and other abductees, all testify to his extraordinary intuitive and healing abilities."[10]

---

6. Randles, 1988, p. 86.
7. Morrow, Jan./Feb. 1993, p. 16.
8. Morrow, Jan./Feb. 1993, p. 17.
9. Mack, 1994, p. 102.
10. Mack, 1994, pp. 238-239.

Another case is that of Jan Whitley of Tujunga Canyon, California, who became obsessed with medicine and a cure for cancer after her UFO abductions.

In 1994, southern California UFO abductee Kim Carlsburg appeared with several other abductees on the television program, *The Other Side*. Kim, who has experienced many terrifying abductions, spoke for herself as well as the entire panel when she said that many of the experiencers she knew, including herself, are psychic healers. As she said, "All of us on this panel are hands-on healers."

Barbara Lamb, who has researched several of the healing cases in this book has also noticed this pattern. In fact, she has a few firsthand cases. As she says, "There is a woman I've been working with, and she is doing Reiki healing. She's been doing that for three or four years now and is very serious about that and really does effectively run energy. And some of her experiences with extraterrestrials involve being taught healings, and with all that being very enhanced. And her son, who is now about six years old...starting at age three...was talking about going on these trips with his space friends at night and riding around in the craft and the things that they'd have him drink, and the strange baths that they'd have him get into and all this stuff he'd never heard anywhere before. She knew he hadn't encountered anything like that in the media or anybody that knew anything. And he and his little cousin, a little girl who is two years younger than he is...have been trained by the extraterrestrials in doing physical healing. And there was one time when the mother, my client, was preparing lunch for him and his little cousin who was visiting for the day. And she cut her hand very badly with a paring knife. And it was bleeding...hurting and stinging.... And it looked like she might have to take herself to the emergency room. She was, of course, concerned.

"And these two kids hopped off their lunch counter stools there in the kitchen and said, 'Oh, we'll heal it. We'll do our space Reiki on you.' And so they went into these movements which she said looked very akin to Tai Chi type of movements. Maybe they were Chi Gong movements for all we know. But anyway, they just healed this cut. She said within a half a minute, the pain really reduced and the blood stopped and it started to congeal and heal up. And I saw her probably about three days later and it was just barely a little line on her hand that I could see had been this big, open, bleeding cut. She could feel the heat coming toward her from them.

"Another woman that I work with...does body work, massage work. And she's only done that for a few years too. She used to work for a corporation and do stuff that was really different. And in some of our regressions we realized that she is learning a lot of energy healing work from the extraterrestrials when she's onboard the craft...she still does some of that energy work with them, learning a lot about the human energy fields."[11]

Some of the experiencers I have worked with have also shown remarkable healing abilities. One lady has recently become a massage therapist and reports that sometimes when she is massaging somebody, she feels the presence of her alien friends helping her to heal her patients. Another lady, Alicia Hansen, who has been healed twice herself, has also proven herself to be a remarkable psychic healer. In 1995, she cured someone of a fractured toe using the technique of laying on of hands. The patient and his doctor were both suitably impressed.

There are many other cases, but by now, the pattern should be clear. Of the many effects that UFOs have upon people, one is definitely the ability to become a healer to cure other people and even animals. It has been reported consistently enough to become a predictable effect.

But what does this pattern mean? Do aliens want people to learn alternative medicine? Are extraterrestrials interested in people who are already natural psychic healers? How do the aliens turn a person into a healer?

Hopefully, cases like these will become the focus of future research so that these questions can be answered more fully.

---

11. Personal files.

KESARA©93'

# Other Miraculous Cures

Miraculous cures have been reported from many sources other than UFOs. People have reported cures from shamans and medicine men of so-called "primitive" cultures. Other cures have been made by religious figures. There are several cases on record in which ghosts and spirits have cured people of illnesses. Furthermore, there are countless numbers of psychics who specialize in healing by "laying on of hands" and other paranormal methods.

Cures by Native American shamans are well-documented. Several well-known cures have been performed by world-famous medicine man Rolling Thunder. Rolling Thunder is a spiritual leader and official spokesman for the Shoshone and Cherokee tribes. He has gained the deserved reputation for being able to cure all kinds of diseases and injuries.

One very well-documented example occurred in April, 1971 at the Menninger Foundation in Council Grove, Kansas. Rolling Thunder was there to give a speech. While there, a young gentleman was discovered to be gravely injured. His leg had been gashed and the wound had become badly infected. There were many modern medical doctors present, several of whom advised that the man be rushed to the hospital.

Rolling Thunder, however, volunteered to perform a healing ritual. By the time everything had been prepared, the man was in considerable pain and no longer able to walk.

Rolling Thunder began the ritual by lighting his peace pipe and passing it to the patient. Also a piece of raw meat was placed in a basin. The purpose of the meat was to hold the negative energies of the injury. Rolling Thunder counseled the patient on his injury, making sure that the healing was justified.

Then he began a "wailing chant" and began sucking the man's wound. After a few moments, he turned and vomited into a basin. This procedure was repeated. Rolling Thunder also placed his hands on the wound, and then took a sacred feather and made passing motions over the wound.

Rolling Thunder announced that the healing was complete and the medical doctors rushed forward to examine the wound. To their utter surprise, the color of the flesh was normal, the swelling had decreased, and the injured man said that he no longer felt any pain. In a few more moments, he was up on his feet and engaged in an active game of Ping-Pong.[1]

Rolling Thunder performs healing rituals on a somewhat regular basis. Each time, he uses the same technique. One ritual that was observed and recorded involved the healing of three patients. One involved a cure for personality problems. Another involved the partial healing of a polio-related condition. And the third patient was cured of tumorous growths in her throat that had ruined her ability to sing.[2]

Another well-known shaman-healer is the Apache scout, Stalking Wolf, who is probably best-known as the teacher of Tom Brown Jr., who now runs the largest wilderness school in the United States. Tom Brown Jr. has also written a series of best-selling books telling of his adventures with Stalking Wolf.

Says Tom Brown Jr., "Grandfather was a tremendous healer.... He had tremendous command of the uses of the plant people, using herbs to cure illnesses that doctors and modern medicine had given up on.... His ability to help people overcome all illnesses bordered on the miraculous."[3]

One of his most miraculous and well-documented cures involved a lady who was stricken with cancer. The lady was in a coma from the cancer which was affecting her entire body. She was expected to die within hours and was allowed to go home to die.

Grandfather Stalking Wolf was asked to perform a healing. Tom Brown Jr. witnessed the entire ritual. As he says, "I saw Grandfather's body begin to vibrate slightly, almost imperceptibly. In the dim room I could see his hands, as if they glowed, and I had to shake my head to make sure I wasn't seeing things. At that moment, like a dull flash

---

1. Boyd, 1974, pp. 14-23.
2. Boyd, 1974, pp. 85-94.
3. Brown, 1988, p. 209.

the old woman's body glowed also, as if illuminated from within, then her body fell back into the original shadow."

Immediately the color returned to the lady's face and she awoke from her coma. Stalking Wolf told the amazed on-lookers, "She will walk within the hour but have no recollection of what has taken place. She will be restored to her full health within seven suns."

When Tom Brown Jr. asked Stalking Wolf how the cure had taken place, Stalking Wolf told him that it was the "life force" and "the spirit-which-moves-through-all-things."

They both saw the lady a few days later, and she was in perfect health. As Tom says, "All traces of illness and coma were gone."[4]

Another well-known shamanistic healer is the late Evelyn Eaton who was taught by a Paiute medicine man and was a member of the Bear tribe. One of Eaton's most dramatic healings involved an injured cat.

Her neighbor's cat had been hit by a car. The cat's back was broken and the cat was totally paralyzed. The vet had been called to put the animal to sleep.

Eaton, however, did not know that the cat was injured. She simply passed her hand softly over the cat's back and squeezed his tail. The cat jumped up, perfectly healed. The pet's owner was totally amazed. Since that incident, Eaton became a well-known psychic healer, until her death in 1983. There are many other well-known shamanistic healers, but the patterns are generally the same.[5]

Another similar type of miraculous cure is known as faith-healing. There are many spiritualist churches in which alleged healings take place on a daily basis. These spiritualist churches are the result of the Spiritualist Movement in the mid-nineteenth century. During this time, there are dozens of healers who cured people of all manners of illnesses.

Today, healings are also coming from channelers, such as Ramtha. J. Z. Knight, who channels Ramtha, was cured of cancer by paranormal means. She was inside a church, and the minister was praying for her recovery when "there was a sudden flash of blue light from the top of the tent. The flash became an electrical flash, almost like lightning and just as bright. The blue lightning streak went verti-

---

4. Brown, 1988, pp. 213-215.
5. Eaton, 1982, p. 10.

cally down, straight through my body." Afterwards, J. Z. was diagnosed as being free of her cancer.

She has also been able to affect miraculous cures on her children.[6]

There are also the well-known religious visitations that have resulted in sacred healing springs. These can be found through-out the world, but the most famous in undoubtedly that of Lourdes, France. A young girl, Bernadette of Lourdes, received a visit by what appeared to be the Virgin Mary. UFO researcher, Jacques Vallée, who noticed the many parallels between religious visitations and UFO events, made a study of the healings which have occurred at Lourdes. One of these healings involves a man who was cured of a severe leg injury after partaking of the water from the Lourdes spring. Another was supposedly cured of blindness. Jacques Vallée has studied the Lourdes healings extensively. He believes that healings have occurred and not just of psychosomatic illnesses. As Vallée says, "The cures performed at Lourdes, however, are not limited to such [psychosomatic] illnesses but extend to such improbable cures as tumors and broken bones."[7]

One well-verified Lourdes cure occurred in 1875 to a Belgian man named Pierre de Rudder. Rudder suffered a compound fracture to his leg, which became infected. Doctors recommended amputation but Rudder refused and begged to go to Lourdes. Before being taken to Lourdes, Dr. Van Hoestenberghe examined the leg and reported, "There was no sign of healing.... The lower part of the leg could be moved in all directions. The heel could be lifted in such a way as to fold the leg in the middle. It could be twisted with the heel in the front and the toes in the back, all these movements being restricted only by the soft tissues."

Rudder was then taken to the spring at Lourdes where he was overcome by a "strange feeling." Rudder suddenly realized he was cured. His wife promptly fainted and Rudder was rushed to a nearby home and examined. As the report reads, "Not only was the wound neatly closed, but the leg had become completely normal again."

Rudder became an instant celebrity. Rudder's own doctor didn't believe what he heard and examined Rudder for himself. The doctor

---

6. Knight, 1987, pp. 250-278, 366-367.
7. Vallée, 1975, p. 158.

allegedly broke into tears and said, "You are completely cured, de Rudder. Your leg is like that of a newborn baby."

De Rudder lived a happy, normal life for twenty-three more years, until he died in 1898 of pneumonia.[8]

In 1917, in Fatima, Portugal, thousands of people witnessed an event that many people consider to be a religious miracle. Three children were given messages from the Virgin Mary, who told them that there would be a display on a certain date. This prediction attracted thousands of witnesses, all of whom saw a glowing disk darting through the clouds. This event is very well-known. What is not as well-known is that there were several miraculous cures affected during the event. As one witness reports, "My mother who had a large tumor in one of her eyes for many years, was cured. The doctors who had attended her said they could not explain such a cure."[9]

There have been several other well-documented cures from these areas. Both events are considered by the Catholic Church to be miraculous in nature.

There are countless other healings that come from religious sources. Conyers, Georgia is now becoming a well-known spot where many miraculous healings of a religious nature have been performed.

The Eastern religions have all kinds of different gods, goddesses and deities whose sole purpose is to assist in healings. These many deities are essentially invoked by the healer while curing a patient.

Of course, there are the accounts of Jesus, who cured one man of blindness, another of lameness and another of leprosy.

Religious figures are only one of the many types of entities who can perform miraculous cures. There has been a recent upsurge in the popularity of angels. Many people are now reporting modern day encounters with angelic beings who save their lives or render invaluable assistance. One little-known case involves a boy who was cured of cancer by an angelic being. Ralph Walker lives in the northeastern United States.

While in high school, he was an avid tennis player. Unfortunately, his playing abilities were hampered by an aching pain in his right arm. He resisted seeing the doctors, hoping that the pain would go away. Instead, the pain became worse. Doctors were called in. Blood tests and X-rays confirmed that Ralph had a malignancy in his upper

---

8. Vallée, 1975, pp. 158-161.
9. Vallée, 1975, p. 150.

right arm. Doctors planned to operate immediately. They told him to give up all expectations of ever playing tennis again.

Ralph's mother, however, was a religious woman. She told Ralph that an angel appeared to her and told her that he would be all right. The operation was scheduled for the next day. That night, Ralph awake and saw a man "wearing a white scrub suit with a stethoscope around his neck, carrying blood pressure equipment." Although it was the middle of the night, Ralph assumed the man was a normal doctor. The man told Ralph that he would be all right and would be able to play tennis again. The man stared into Ralph's eyes and Ralph was overcome with a feeling of "warmth that seemed to permeate both arms and legs and then his whole body."

The next morning, Ralph mentioned the doctor. His doctors denied having visited him and said he must have dreamed it. Ralph denied this and said that the doctor was "like any doctor. Dressed in white. I sure didn't dream him."

Ralph was taken to the hospital where last-minute X-rays were taken. It was then discovered that the cancer was gone. As Ralph's doctor said, "I have to say that nothing like that has ever happened in my twenty-three years of practice.... I mean, Mr. Walker, that these new X-rays show absolutely no trace of any cancer. If more X-rays and tests confirm it, there is no need for an operation."

Needless to say, Ralph is fine. He no longer believes that the man who came into his bedroom that night was a doctor. He believes he was healed by a guardian angel.[10]

There are also countless psychic healers. Without a doubt, one of the most famous is the late Edgar Cayce, who was able to diagnose and prescribe cures for literally thousands of illnesses. Cayce always performed his healings in a sleep-like trance state. He usually prescribed unusual remedies. Because of his unparalleled success in such healing cases, he has become world famous as a psychic healer.

I personally have investigated an account of a spontaneous psychic healing done by a lady whom I shall call Diane Robinson. Robinson has had extensive UFO encounters as well as paranormal experiences of several types. One of these experiences includes a healing. She had recently heard on television that healing could be done by simply laying one's hands on the patient. So when Diane

---

10. Fearheiley, 1993, pp. 54-74.

Robinson's infant daughter suffered one of her severe asthma attacks, Diane decided to attempt a cure.

Usually, the asthma attacks were severe enough to warrant a hospital visit and Diane could tell that this time was no different. But she laid her hands on her child and prayed that a healing would take place. Diane reports that a strange, buzzing warmth filled her hands. It scared her and she jerked her hands away. When she examined her daughter, she was amazed to see that the asthma attack had stopped. Her daughter was breathing with perfect ease.[11]

Many people who have had near-death experiences are also reporting being healed of injuries and illnesses as a direct result of their experiences. I investigated a case involving a young woman who had been diagnosed with terminal cancer. During her hospital stay, she had a classic near-death experience. Upon awakening back in the hospital, all traces of the disease were gone.

Even more astonishing are "mixed motif" cases which involve elements of both a near-death experience and a UFO experience. The two experiences have a number of striking parallels, and as it turns out, a small minority of cases could be labeled under either category. Kenneth Ring Ph.D. and Raymond Fowler have made studies of the parallels of the near-death experience and the UFO experience.[12] In his book, *The Omega Project,* Ring presents a very interesting "mixed motif" case involving a physical healing.

In 1977, day-care center operator Beryl Hendricks of New York State discovered a tumor in her breast. Doctors promptly removed it and to Beryl's relief, the tumor was benign.

One year later, Beryl was shocked to discover another tumor about the size of a golf ball in her breast. She resolved to call her doctor the next day and went downstairs and joined her husband on the couch. Suddenly, she fell unconscious. Her husband was unable to feel a pulse.

Meanwhile, Beryl found herself apparently out of her body, aboard what appeared to be a spacecraft. Says Beryl, "The next thing I remember I was looking out of a round window and seeing the blackest blackness with tiny white sparkles...." Beryl turned around and suddenly realized she had been placed on "some kind of operating table" surrounded by several "thin, tall figures." A bright light

---

11. Personal files.
12.

was shining down upon her. The figures spoke telepathically, saying, "Look and see—it is gone." They also told her to "follow her husband," evidently because they had been having marital difficulties.

Beryl then found herself racing out of her body back to her still form on the couch. She woke up and began vomiting. Two hours had passed. Almost immediately, Beryl made the astonishing discovery that "the lump was gone—totally." Ten years later, Beryl continues to enjoy excellent health.

Was Beryl healed as a result of a near-death experience or a UFO encounter? As Kenneth Ring says, "When you read it, please decide into which of our two categories it should be placed."[13]

There are also consistent reports of healings after being struck by lightning. One recent example of a lightning-strike cure is that of Mary Clamser of Oklahoma City, Oklahoma. When Mary was nineteen years old, she was diagnosed with severe multiple sclerosis. She suffered with the disease for over twenty years, and her condition gradually worsened. However, on August 14, 1994, she was struck by lightning while in her home. Although she suffered burns and other harmful side-effects, she instantly began a miraculous recovery from her multiple sclerosis. Feeling came back into her legs, and less than a month later, she began walking again. As Clamser says, "I knew something had changed. I was so overwhelmed that I started to cry." Today, she continues her miraculous recovery.[14]

Most of the above cures are spiritual or psychic in nature. Most of the alien cures, however, do not seem to be psychic in nature. More often in fact, they appear to be technological in nature.

Many doctors in the United States use technological methods also. The number of success stories by modern doctors is truly incredible. It may be for this reason that we find so many parallels between how modern human doctors and alien doctors perform cures. We both use invasive surgery. We both have highly technological surgical instruments. We both use medicines and injections.

The parallels are so striking that it is difficult to imagine how so many UFO witnesses who have been healed could be lying about their experiences. The stories are simply too similar to each other and to modern accounts of surgery.

---

13. Ring, 1992, p. 109.
14. Paxton, 1995, pp. 79-80.

The main difference between modern human cures and UFO cures is that UFO cures involve superior technologies. The medicine is more powerful. The surgery is more effective. The instruments are able to perform more functions.

However, what do all these healings mean? Are there any patterns? What's the most common type of cure?

A statistical scientific examination of the data will answer all these questions and more. In the next chapter, we will learn the many patterns behind the UFO healings.

KESARA

# A Chronology and Statistical Analysis

A scientific study of any type demands a statistical analysis of the reported data. To assist in the analysis, I have organized the UFO healings into chronological order. Each case has been assigned a number respective to the date the healing was performed. Those cases without dates have been placed at the end of the chronological list.

The information that will be analyzed includes the date of the healing, the location of the healing, the person who received the healing, the disease/illness/condition healed and the method or manner of healing. Finally, each case will be indicated as involving either a contactee or abductee. If neither is indicated, no onboard experience took place.

What follows is the longest list of UFO healings ever assembled.

CASE #001. Mid-1930s. Location Unknown, USA. Anonymous male. Tuberculosis. Operation onboard UFO. Abductee.

CASE #002. 1937. Switzerland. Eduard Meier. Pneumonia. Visit in bedroom. Contactee.

CASE #003. 1942. Philadelphia, PA. Edward Carlos. Pneumonia. Hospital visit. Abductee/contactee.

CASE #004. April 1945. Okinawa. Howard Menger. Blindness. Hospital visit. Contactee.

CASE #005. 1940s. Southern California. Anonymous male. Hole in heart. Energy onboard UFO. Abductee.

CASE #006. Early 1950s. New Jersey. Richard Rylka. Ear Infection. Hospital visit with alien mind power. Contactee.

CASE #007. Early 1950s. Southern California. Anonymous female. Flesh wounds. Operation onboard UFO. Abductee.

CASE #008. April 1952. New Jersey. Ellen Crystall. Enlarged stomach. Close association with a UFO.

CASE #009. July 1952. Agawam, MA. Marianne C. Shenefield. Given "psychic sight." Close association with a UFO. Abductee.

CASE #010. Mid-1950s. Southern California. Several anonymous people. Skin cancer. Alien prescribed-cure involving vinegar applied to skin by human physician.

CASE #011. October 25, 1957. Petropolis, Brazil. Anonymous young female. Cancer. House visit with operation, light beams and pills.

CASE #012. 1957-1958. West Midlands, England. Anonymous extraterrestrial. Finger burn. Jelly-like spray applied.

CASE #013. Summer 1959. Pleasanton, TX. Two roosters. Body injuries. Light beam from UFO.

CASE #014. September 27, 1959. Coos Bay, OR. Leo Bartsch. Numbness in hand. Close association with UFO.

CASE #015. 1950s. Eastern Transvaal, South Africa. Ann Grevler. Incision on leg. Alien mind power. Contactee.

CASE #016. 1962. Location unknown. Alice Haggerty. Diphtheria. Beams onboard UFO. Abductee.

CASE #017. September 3, 1965. Damon, TX. Robert W. Goode. Flesh wound on finger. Beam from UFO.

CASE #018. September 1966. Durhan, England. Fred White. Hole in lung. Hospital visit with alien instrument and mind power.

CASE #019. December 7, 1967. Denmark. Hans Lauritzen. Liver hepatitis. Close association with UFO.

CASE #020. 1967. Lima, Peru. Ludwig F. Pallman. Kidney disease. Hospital visit with alien mind power and pills. Contactee.

CASE #021. November 2, 1968. French Alps. "Doctor X." Partial paralysis and flesh wound on ankle. Beam from UFO.

CASE #022. December 9, 1968. Peru. Anonymous male. Myopia. Beam from UFO.

CASE #023. December 9, 1968. Peru. Anonymous male. Rheumatism. Beam from UFO.

CASE #024. 1960s. New Jersey. Richard Rylka. Crushed finger. Hospital visit and alien mind power. Contactee.

CASE #025. 1960s. Kentucky. Jerry Wills. Fever. House visit and injections. Contactee.

CASE #026. 1970. New Jersey. Richard Rylka. Body injuries from auto accident. Method of cure unknown. Contactee.

CASE #027. 1971. Southern California. Fred Bell. Burn on hand. Alien instrument. Contactee.

CASE #028. 1972. Colorado. Anonymous female. Infertility. House visit. Abductee.

CASE #029. December 30, 1972. Argentina. Ventura Maceiras. New teeth. Beam from UFO.

CASE #030. March 1973. United States. Olga Adler. Chronic back pain. House visit and alien instrument. Abductee.

CASE #031. October 1973. Mid-western United States. Deborah Roach. Illness not described. Operation onboard UFO. Abductee.

CASE #032. 1973. Marseilles, France. Denise B. Throat cancer. Beam from a UFO. Abductee.

CASE #033. January 1974. Las Vegas, NV. Frank E. Stranges. Body injuries. Beam onboard UFO. Contactee.

CASE #034. July 8, 1974. Brooklyn, New York. Brandon Blackman. Flesh wound on finger. Close association with a UFO.

CASE #035. May 10, 1975. Florence, KY. Chuck Doyle. Head cold. Beam from UFO.

CASE #036. 1975. Arizona. Helen X. Cancer. Operation onboard UFO. Abductee.

CASE #037. April 3, 1976. Switzerland. Eduard Meier. Fractured rib. Alien instrument. Contactee.

CASE #038. 1976. Medicine Bow National Monument, WY. Carl Higdon. Kidney stones. Close association with UFO. Abductee.

CASE #039. 1976. Medicine Bow National Monument, WY. Carl Higdon. Tubercular scar on lung. Close association with UFO. Abductee.

CASE #040. Mid-1970s. Florida. Anthony Champlain. Hand injury. Beam aboard UFO. Contactee.

CASE #041. July 1978. Switzerland. Eduard Meier. Pneumonia. Operation onboard UFO. Contactee.

CASE #042. August 30, 1979. Eustache, Quebec, Canada. Jean Cyr. Multiple Sclerosis. Close association with UFO.

CASE #043. 1979. Sturgeon Bay, WI. Anonymous young female. Muscular dystrophy. Beam onboard UFO. Contactee.

CASE #044. 1970s. Southern California. James X. Hematoma on head. Operation and beam onboard UFO. Abductee.

CASE #045. August 1980. San Juan, Puerto Rico. Hector Vasquez. Kidney stones. Close association with UFO. Friend of contactee.

CASE #046. November 19, 1980. Longmont, CO. Michael X. Skin cancer. Close association with UFO. Abductee.

CASE #047. November 28, 1980. West Yorkshire, England. Alan Godfrey. Infertility. Close association with UFO. Abductee.

CASE #048. November 1981. Hubbard, OR. Denise Twiggs. Caesarean section. Hospital visit. Contactee.

CASE #049. August 1982. N. E. Iowa. Dagmar R. Skin mole. Beam onboard UFO. Abductee.

CASE #050. 1984. Botucatu, Brazil. Joao Vicente. Cerebral aneurysm. Alien salves applied to skin. Friend of contactee.

CASE #051. 1986. La Jolla, CA. Richard T. Lower-body paralysis. Method of cure unknown. Abductee.

CASE #052. Mid-1980s. Philadelphia, PA. Edward Carlos. Skin cancer. Beam of light. Abductee/contactee.

CASE #053. 1987. Atlanta, GA. Alicia Hansen. Body injuries. Beams of light onboard UFO. Contactee.

CASE #054. March 20, 1988. Richland Center, WI. John Salter, Jr. Integumentary system. Onboard operation with injections. Abductee.

CASE #055. May 1988. Sebago Cabins Recreation Park, NY. Anonymous male. Swollen and painful legs. house visit. Friend of contactee.

CASE #056. October 1988. New York. Anonymous female. Incision. Operation onboard UFO. Abductee.

CASE #057. December 1988. Watsonville, CA. Anne DeSoto. Arthritis of hip. Operation onboard UFO. Abductee.

CASE #058. July 5, 1989. Newark, NJ. Richard Rylka. Tumor on forehead. Operation onboard UFO. Contactee.

CASE #059. July 14, 1989. Hubbard, OR. Bert Twiggs. Severe cold. house visit with injections. Contactee.

CASE #060. August 7, 1989. Pensacola, FL. Katharina Wilson. Heart injuries due to lightning strike. Operation onboard UFO. Abductee.

CASE #061. September 4, 1989. Amsterdam, Netherlands. Jan DeGroot. Wart on neck. Close association with UFO.

CASE #062. October 15, 1989. Hubbard, OR. Bert Twiggs. Back injury. Alien instrument onboard UFO. Contactee.

CASE #063. October 15, 1989. Hubbard, OR. Stacey Twiggs. Asthma. Alien medicine. Contactee.

CASE #064. December 1989. Hubbard, OR. Christopher Twiggs. Sty in eye. Operation onboard UFO. Contactee.

CASE #065. December 15, 1989. Mississippi. Mae X. Arthritis. Instrument aboard UFO. Abductee.

CASE #066. 1989. Los Angeles, CA. Licia Davidson. Colon cancer. Operation onboard UFO. Abductee.

CASE #067. 1989. Deming, NM. Anonymous female. Dizziness (caused by medication). Method of cure unknown (missing time). Abductee.

CASE #068. April 15, 1990. Island of Iona. Edward Carlos. Arterial blockage in heart. Operation onboard UFO. Abductee/contactee.

CASE #069. 1990s. Finland. Anonymous Male. Congenital liver disease. Method of cure unknown. Abductee.

CASE #070. 1991. Claremont, CA. Anonymous female. Back pain. Beam from UFO.

CASE #071. 1991. Atlanta, GA. Alicia Hansen. Cyst. Operation onboard UFO. Contactee.

CASE #072. February 1992. Topanga Canyon, CA. Michelline. Flesh wound. Mind power taught be extraterrestrials.

CASE #073. December 1992. Virginia. Beth Collings. Improved eyesight. Method of cure unknown. Abductee.

CASE #074. 1993. Virginia. Anna Jamerson. Improved eyesight. Method of cure unknown. Abductee.

CASE #075. September 1993. Kokomo, IN. Debbie Jordan (aka Kathie Davis). Tumors. Operation onboard UFO. Abductee.

CASE #076. June 1994. Harrison County, WV. Daniel D. Impacted wisdom teeth. House visit. Abductee.

CASE #077. September 12, 1994. USA. "David X." Abdominal/intestinal illness. Operation onboard UFO. Abductee.

CASE #078. December 13, 1994. Southern California. Morgana Van Klausens. Cyst. Close association with UFO. Abductee.

NOTE: The following cases do not have the dates of the healings.

CASE #079. S. W. United States. Anonymous male. Burn on hand. Alien spray. Abductee.

CASE #080. Santa Clara, CA. Ted X. Angioma. Operation with beams onboard UFO. Abductee.

CASE #081. Chicago, IL. Doriel X. Infertility. Close association with UFO.

CASE #082. East Kilbride, England. Anonymous male. Tumor in chest. Close association with UFO. Abductee.

CASE #083. Toronto, Canada. John Adams. Polio. house visit and operation. Contactee.

CASE #084. Hungary. Anonymous female. Cancer. Operation onboard UFO. Contactee.

CASE #085. Peru. Mr. X. Stomach cancer. Beam of light from UFO.

CASE #086. Newark, NJ. Richard Rylka. Polio. house visit with alien instrument. Contactee.

CASE #087. Georgia, Russia. Anonymous male. Heart attack. house visit. No prior association.

CASE #088. Chicago, IL. Ann X. Sinusitis. Instrument onboard UFO. Abductee.

CASE #089. Midwestern USA. Eddie X. Color blindness. Operation onboard UFO. Abductee.

CASE #090. [no location given] Star Traveler. Head injury. House visit. Abductee.

CASE #091. Hubbard, OR. Denise Twiggs. Colitis. Operation onboard UFO. Contactee.

CASE #092. Hubbard, OR. Bert Twiggs. Lung Cancer. Operation onboard UFO. Contactee.

CASE #093. Tblisi, Georgia, Russia. Anonymous male. Knee injury. Beam from UFO.

CASE #094. Tblisi, Georgia, Russia. Anonymous male. Heart condition. Operation.

CASE #095. Tblisi, Georgia, Russia. Anonymous young girl. Diabetes. Alien instrument onboard UFO. Abductee.

CASE #096. Tblisi, Georgia, Russia. Anonymous male. Chronic back pain. house visit and operation.

CASE #097. California. Linda X. Yeast infection. Operation and salve onboard UFO. Abductee.

CASE #098. California. Linda X. Cancer. Operation onboard UFO. Abductee.

CASE #099. California. Linda X. Liver infection. Salve applied onboard UFO. Abductee.

CASE #100. California. Sherry X. Yeast infection. Operation and salve onboard UFO. Abductee.

CASE #101. Western USA. Anonymous male. Stomach ulcer. Alien instrument onboard UFO. Abductee.

CASE #102. Western USA. Anonymous young female. Intestinal worms. Injection onboard UFO. Abductee.

CASE #103. Western USA. Anonymous young male. Hip problems. Alien instrument onboard UFO. Abductee.

CASE #104. Western USA. Anonymous male. Bursitis. Alien instrument onboard UFO. Abductee.

CASE #105. Western USA. Anonymous male. Tumor on head. Alien instrument onboard UFO. Abductee.

There is a total of 105 cures, however, several cures were not included in this book because of lack of data. The above cures are by no means the total number of all healings, but they do represent an accurate cross-section. Also, with 105 reported cases, the data base is large enough to conduct a fairly accurate statistical analysis.

The dates of the cures range from the mid-1930s to 1994. Twenty-five of the cases provided no date. Statistically, the dates of the cures are as follows:

1930s: 2/105 (2%)
1940s: 3/105 (3%)
1950s: 10/105 (10%)
1960s: 10/105 (10%)
1970s: 20/105 (19%)
1980s: 23/105 (21%)
1990s: 10/105 (10%)
Unknown: 27/105 (26%)

The locations come from all over the world. The preponderance of accounts from the United States can be explained by the fact that this study was conducted from the United States. The other reports indicate what many other UFO studies have already shown: that UFOs are a global phenomenon. What follows is a list of the locations of the cures.

Argentina: 1 (1%)
Brazil: 2 (2%)
Canada: 2 (2%)
Denmark: 1 (1%)
Finland: 1 (1%)
England: 4 (4%)
France: 2 (2%)
Hungary: 1 (1%)
Iona, Island of: 1 (1%)
Netherlands: 1 (1%)
Okinawa: 1 (1%)
Peru: 4 (4%)
Puerto Rico: 1 (1%)
Russia: 5 (5%)
South Africa: 1 (1%)
Switzerland: 3 (3%)
United States: 70 (67%) (AZ, CA[15], CO[2], FL[2], GA [2], IL[2], IN, IO, KY[2], MA, MS, NJ[6], NM, NV, NY[3], OR[8], PA[2], TX[2], VA[2], WI[2], WV, WY[2].)
Unknown: 3 (2%)
Vietnam, North. 1 (1%)

An examination of the people cured of UFOs shows some strange patterns. Of the 105 cures, 64 (64%) involve men, 38 (36%) involve women. There is also one case involving a male alien, one case involving two roosters, and one case involving several anonymous patients.

The fact that there are nearly twice as many men cured by UFOs than women may seem inaccurate, however, for some reason, there are more male contactees (who have gone public!) than female contactees. And as we have seen, contactees receive many of the cures.

But surprisingly, the data shows that abductees receive many more cures than do contactees!

In terms of abductee versus contactee, 47 (45%) of the cases involve abductees, 29 (28%) involve contactees. Three (3%) involve friends of contactees. Twenty-four (22%) involve people who have no prior association with UFOs. One case involves an alien and one involves animals.

The cures have occurred to people from ages of pre-birth to upwards over 60 years of age.

The cures themselves also follow certain patterns.

Of the 105 cures, 24 (22%) involve healings of flesh wounds and injuries, 16 (15%) involve healings of minor illnesses and ailments, 7 (7%) involve eye cures, 4 (4%) involve cures of the integumentary system, 6 (6%) involve kidney and liver cures, 6 (6%) involve lung cures, 31 (30%) involve cures of serious and/or chronic illnesses/diseases, and 11 (10%) involve cancer cures.

What follows is a list of all the reported cures, in alphabetical order. Unless indicated, there is only one case per cure:

Aneurysm (cerebral [2])
Angioma
Arthritis [2]
Asthma
Back pain [4]
Blindness & Color Blindness
Body injuries [5]
Burn (finger, hand [2])
Bursitis
Cancer (breast, colon, skin [3], stomach [2], throat, full-body [2], lung)
Candidiasis (yeast infection) [2]
Caesarean section
Cold (common [2])
Colitis
Diabetes
Diphtheria
Dizziness
Ear infection
Fever
Flesh wounds (finger [3], ankle, hand, legs, genitals)
Head injury
Heart condition [4]
Hematoma (head)
Hip problem
Incision [2]

Infertility [3]
Integumentary system (improvements)
Intestinal worms
Intestinal disease
Kidney (disease, stones [2], failure)
Knee injury
Liver disease [3]
Lung (hole, tubercular scar)
Mole
Multiple Sclerosis
Muscular Dystrophy
Myopia/Poor eyesight [3]
Numbness (hand)
Paralysis [2]
Pneumonia [3]
Polio [2]
Rheumatism and Rheumatoid Arthritis
Rib (fractured)
Sinusitis
Stomach problem
Sty (in eye)
Teeth (new set, impacted wisdom)
Tuberculosis
Tumor (chest, head [2], abdomen [2], breast)
Ulcer (stomach)
Unknown
Wart

This impressive list of cures makes it very clear that no illness is too small or too great to be cured by aliens. It doesn't matter if the condition is irritating, chronic or fatal—each has been cured. The most common cure of a specific type is cancer.

The methods of cures fall into four main categories. There are cures performed onboard UFOs, in homes, in hospitals and outside. The cures involve beams of light, surgical operations, strange instruments, pills and salves, close association with a UFO and alien mind power. Of course, some of the cures involve various combinations of the above cures.

Twenty-three (21%) of the cases involve some type of light beam from a UFO. In 7 (7%) of these cases, the light beam was experienced during an onboard experience. At least 27 (26%) of the cases involve surgical operations onboard a UFO. Three (3%) involve surgical operations inside the witnesses' homes. Fourteen (13%) of the cases involve close association with a UFO. Fourteen (13%) of the cases involve alien instruments held over the body, not including those instruments that were used in surgical operations. Nine (9%) of the cases involve pills, salves or medication. Five (5%) of the cases involve injections. Five (5%) of the cases involve alien mind power.

Of the 105 healings, 52 (49%) occurred aboard a UFO, 21 (20%) occurred outside, at least 22 (22%) of the cures were performed in the witness's home, and at least 9 (9%) of the cures were performed in a hospital.

There are some other surprising patterns. Many of the cures happened to just a few people. The total number of people cured is 87, not including animals. A surprising 28 (27%) of the 105 cures involve only 12 people. In other words, 11% of the cures involve people who have been cured more than once. Richard Rylka received at least five (5%) cures and the Twiggs family, which has five members, received seven (7%) cures. Abductees/contactees, Linda X., Edward Carlos and Eduard Meier each received three (3%) cures. An anonymous Peruvian customs official, an anonymous French doctor, Alicia Hansen and Carl Higdon all received two cures (2%) in one encounter.

There are at least three cases in which the witnesses were not cured themselves, but saw other people being cured aboard the UFO. Most of the people who are cured by aliens have no relationship with other people who have been cured. However, there are exceptions. Two women who received a healing of their eyesight are both friends and co-workers. Two other women who both received cures are sisters. There is also one family of five in which each of the members have received at least one cure.

These statistics tell us many things. They tell us that cures happen to all types of people all over the world, regardless of age, sex, educational level, religion...etc. This would indicate that extraterrestrials are definitely not prejudiced.

The cures cover a wide range of diseases and use a variety of the methods. Many of the illnesses have been labeled chronic and/or incurable by modern doctors. The severity of some of the diseases cured clearly shows that the aliens have advanced well beyond humans in terms of medical technology. The cures are done with very advanced methods and technologies and have been occurring for over fifty years, with a slow and steady increase in cases.

Many of the cases were verified by doctors and other officials, and most of the cases were researched by well known investigators. The cases are so numerous and widespread that they can no longer be ignored.

The patterns are clear. For reasons that remain unknown, aliens are very interested in human beings.

# Epilogue

Even as this book is being written, new healing accounts continue to be reported. In the first annual Gulf Breeze Conference, UFO investigator John Carpenter disclosed the details concerning one of his strangest UFO abduction cases. The case involved a young girl who experienced an abduction during which the lower half of her body was "cut away, taken out of the room, and a new lower half was brought in and attached to her!" Today the 16-year-old girl is allegedly "unusually proportioned."[1]

Another recent case involves a woman named Rayna, her daughter and two others. All four experienced a simultaneous missing time abduction in Sedona, Arizona. The mother used hypnotic regression and recalled a typical abduction experience. Her daughter on the other hand was too young to undergo hypnosis. Nevertheless, she showed signs of remembering her experience. As her mother says, "My daughter was two when she started having experiences. She used to take dolls and put them under a lampshade and say, 'I'm healing them.' She talked about white men who came in through the window...."[2]

These two cases are typical of how UFO experiences can be interpreted in extreme ways. The first case might cause somebody to take a negative view towards abductions in that it appears that the aliens are simply performing bizarre experiments. The second case, however, gives the impression that UFO abductions are beneficial, healing events.

Which is the correct interpretation? Are abductions for people's own good? Are aliens good or bad?

---

1. Norris, Vol. 1, No. 12, p. 7.
2. Coats, Feb. 1994, pp. 8-9.

Writing this book has forced me to examine these types of questions. When I started the investigation, I had no idea how radically it would change my beliefs about the UFO phenomenon.

Like many UFO investigators, I entered the field with a prejudice against contactees. At first I simply ignored their stories. But when they became too numerous, consistent and credible, I found myself in an unenviable position. Was it really objective and scientific to continue to ignore their stories? Obviously it wasn't. So if I wanted to be true to myself, I had to at least examine the evidence.

Some of the cases in this book have generated controversy. Some UFO investigators regard the contactee phenomenon with disgust and disdain. If the aliens are friendly, they don't want to hear about it. Although some cases have been more fully investigated than others, the majority of them were verified in some way by doctors as well as investigated by well known figures in the field of UFOs.

Furthermore, I found the evidence for contactees to be as convincing as the evidence for abductees. Actually, contactees usually presented more and better evidence than the typical abductee, although the contactee stories usually had a higher level of strangeness. And when I started uncovering first-hand contactee cases of my own, I knew the peril I was taking by covering up these stories. Covering up evidence has never helped solve any mysteries. To solve the UFO puzzle, we need all the pieces. And so I began to examine the contactee accounts.

The contactee accounts are wealthy with healings, but as we have seen, there are actually more abductees who have been healed than contactees, and just as many people with no prior association with UFOs.

Researching and writing this book has shown me that the term abductee is often misleading, and that many abductees fit better under the category of contactee. The line between the two is much thinner than I ever realized and may sometimes depend upon how the witness interprets the experience. Many abductees maintain that the experience was against their will and very frightening. Their choice to label themselves as abductees is, of course, quite understandable. However, a person might have a nearly identical experience as an abductee, but would rather be labeled as a contactee simply because they found that the experience was not frightening, but rather beneficial. Ufology has suffered semantic disaster before with the term "UFO" which can mean a number of different things, from an alien spaceship to virtually anything else flying in the sky. It seems that we are struggling to find the right single word that will define a UFO experience a label someone who has had an onboard encounter. Such a word, of course, does not exist.

Probably the greatest effect this book has had concerns how I view the UFO phenomenon. I used to be a skeptic and was sure that UFOs were figments of people's imaginations. After examining the accounts, I thought that perhaps UFOs were explorers or tourists. More accounts changed my opinion of them to scientists. And when the lurid abduction accounts mounted to a frenzy, I found myself beginning to think of UFOs as alien torture chambers. While writing this book, however, I came to view the UFOs more like floating hospitals.

Now, however, I have come to a new conclusion. UFOs can be put under many labels and can be interpreted in a variety of ways. Belief in one theory or another proves nothing. The best way to view the UFO phenomenon is with objectivity.

I am not convinced that abductions or contacts are for our own good or for evil. I do not believe the aliens are benevolent space brothers any more than I believe they are evil invaders. I am aware that some people are affected negatively by abductions. However, I am also aware of many positive effects. After conducting the research for this book, I found myself unable to escape the conclusion that the only way to answer questions about UFO hostility versus benevolence is to remain objective and let the witnesses speak for themselves.

Some people believe that the aliens are here for their own selfish gains. Others believe that aliens are friendly in the true sense of the word—that they care not only for people's physical health, but for their mental health, and for the state of the world. In the abductee/contactee community, UFO benevolence versus hostility remains a hotly contested issue.

Belief, however, proves nothing. The fact that there are differing opinions indicates that our understanding of UFOs remains incomplete. It doesn't matter what people believe because the truth has a way of taking care of itself.

The evidence shows that people are being cured for reasons beyond just testing and experimentation. The fact that some people have had their health consistently maintained by extraterrestrials and that they have a mission, shows that the aliens are actually practicing medicine and are not just experimenting. This implies that the aliens have the knowledge to cure all humanity.

Yet, the aliens are not curing everybody. Whitley Strieber writes in his latest book, *Breakthrough*, that he continually gets letters from readers requesting that he ask the aliens to cure them of disease. As Strieber says, "Many's the night I'd spent praying for the desperate people who have sent me letters. 'Please, Mr. Strieber, I have stage-three Hodgkins and three kids and no husband, please get them to save me.' 'My friend has AIDS, he's suffering in agony, call the visitors, get them to come.' 'I need medicine from the other world or I am going to die.'"[3]

Needless to say, these people are not being cured. However, other cures are still taking place. In his book, Strieber reports that he met "a fascinating witness from California who'd apparently had a cure in connection with her contact experiences."[4]

Each time I speak on this subject, at least one and sometimes several people come up to me to report their own cures. One lady told me that she was cured of deafness as a child, and she had the laser scars behind her ears to prove it! Two different doctors had asked her when she had surgery on her ears. However, she has never had ear surgery, at least not by human doctors.

Another fascinating account comes from a lady who was cleaning her bathtub using a deadly combination of Limeaway and Bleach. When she turned on the water, a lethal gas was created which the lady inhaled. She rushed herself to the hospital where emergency doctors treated her. Her condition, however, did not improve and doctors thought she would die. In fact, the odor of the gas was so strong that it was making the hospital personnel physically ill. To make a truly long story short, about one week after the accident, the lady was taken inside a UFO where grey-type beings healed her using lights and sounds. Needless to say, today she is healthy.

Yet another lady reports that she was cured of blindness and that just recently her opthamologist told her in no uncertain terms that she should not be able to see, even though she can.

Other cases are coming out in the literature. Take the following case, reported by Colorado UFO investigator Christopher O'Brien, involving a cure of cancer.

---

3. Strieber, 1995, p. 161.
4. Strieber, 1995, p. 162.

CASE: Barbara Benara (pseudonym) of the san Luis Valley in Colorado had her first fully conscious encounter with beings she calls "the little brothers" in 1978, one week after she had been diagnosed with ovarian cancer. She went to bed and woke up to find herself "lying on a soft table in a white, circular room. I could only move my head. I don't recall how long I lay there, but I felt peaceful."

Barbara then heard a voice which told her that she needed help. She asked to see who was speaking. As she says, "There were four of them, all identical. They were three to four feet tall, ivory white, and had large, almond-shaped eyes." A slightly taller being stepped forward. Barbara, like many other people, was hypnotized by the alien's eyes. It was then that Barbara underwent an apparent operation to cure her of her cancer.

Says Barbara, "I don't remember how I got there, but somehow I found myself immersed in an L-shaped tank. A long, segmented arm came out of a console next to the tank and inserted itself into me. It wasn't painful, but it didn't feel comfortable. I had to relax and trust the voice that kept telling me not to be afraid."

Barbara woke up tired, but otherwise unharmed. Four days later, she returned to her doctor and demanded that he give her the same tests to determine the status of her cancer. Amazingly, the tests showed that Barbara was free of cancer. As she says, "They cured me of ovarian cancer."[5]

I continue to uncover more cures: sinusitis, a rash, an earache, and one case involving a cure of Lupus. As can be seen, there is no shortage of cures. They have been going on for decades and show no signs of stopping.

If the aliens have the power to cure humanity and they are benevolent, then why don't they help us? It is this issue that has become problematic in the UFO field. Why haven't the aliens showed themselves and landed on the White House lawn? If aliens were here to cure us of our diseases, why are diseases like AIDS and cancer still killing so many people?

Of course, any answer would be pure speculation. However, UFOs have always contacted humanity in a covert manner. There has never been full, open, official contact with the public at large. The obvious reason for this is because of the slow spiritual growth of humanity. The extreme prejudice and hatred displayed by people of all

5. O'Brien, 1996, pp. 36-39.

nations—people who will go so far as to kill to defend their prejudices, could be the single greatest obstacle that is preventing open contact with UFOs. With so many warring nations, the UFOs may be concerned that humanity's prejudice against those who are "different" would be too dangerous for open contact.

And yet, UFOs, without a doubt, are more powerful than humanity. They represent a more advanced technology. And since their technology is more advanced, the chances are good that they are more spiritually and ethically advanced. Logically, the bell curve would dictate that some aliens must be friendly, therefore, they would then attempt to help humanity. It would logically follow that there would be some limited healing activity.

Of course, the aliens are wiser than to cure all humanity's diseases at once. Imagine the consequences. A sudden cessation of all diseases would result in over population and starvation in many areas of the world. Just think what would happen if our military leaders were given healing technology. The ability to cure flesh wounds in seconds would only give warring countries a never-ending supply of soldiers. In all likelihood, the consequences would be even worse.

Also, being handed the answers would not help humanity to be independent and capable of confronting future problems.

The aliens, however, continue to make regular appearances throughout the world. The effect has been to cause a widespread and growing belief in the existence of UFOs. It could be that the UFOs do intend to make open contact in the future, but that humanity must first change.

For the first time, we have matured enough to realize that we are not alone. Although progress is slow, we are becoming more spiritually oriented and more technologically advanced. Although it is taking a long time, we are growing up.

Hopefully, the future will bring about open contact with UFOs and humanity. All life in the universe shares a common heritage of consciousness. We all come from the same Universe. And we all must learn to live in the same universe, together.

Researching this book has proved to me that ultimately, extraterrestrials are people very much like humans, only different. Some are here as scientists. Some are here as explorers. Some are here as doctors. Just like people, they can be categorized under any one of a number of labels. And just like people, some have good intentions, and some may not. But they are all still beings who are living out their lives according to some grand plan.

Hopefully, this book will help further the understanding of all these beings that our visiting our planet.

*Preston E. Dennett*

# Biographical Sketch

Preston Dennett received his Bachelor of Arts degree in English from California State University of Northridge. He first began studying UFOs in 1986 when he discovered that his friends and family were having dramatic encounters. Since then he has interviewed hundreds of UFO witnesses and investigated cases of virtually every type. He is a field investigator for the Mutual UFO Network (MUFON), Seguin, Texas. He is also involved in several other organizations including Citizens against UFO Secrecy (CAUS), Center for the Study of Extraterrestrial Research and the UFO Contact Center International (UFOCCI). UFO cases have been referred to him by the police, he has appeared on *Encounters: The Hidden Truth*, and he has consulted for *Sightings*. He lectures across the United States on the subject of UFOs, and he currently resides in southern California.

Preston Dennett has had over 50 published articles since 1986 in various UFO publications. In the last year, Dennett has published articles in *Messenger, UFO Universe, Unsolved UFO Sightings, UFO Encounters* and *Contact Forum*. Other books by Preston Dennett are *One in Forty: The UFO Epidemic* (Nova Science Publishers, Inc., 1995), *Contactee, The Great Moon Mystery* and *The Topanga Canyon Close Encounters*.

# Sources

**Books:**

Bell, Fred. (as told to Brad Steiger) *The Promise*. New Brunswick, NJ: Inner Light Publications, 1991.

Blum, Ralph and Judy Blum. *Beyond Earth: Man's Contact with UFOs*. New York: Bantam Books, 1974.

Bowen, Charles, ed. *The Humanoids: A Survey of Worldwide Reports of Landings of Unconventional Aerial Objects and Their Occupants*. Chicago, IL: Henry Regnery Company, 1969.

Boyd, Doug. *Rolling Thunder*. New York: Delta Books/Dell Publishing Co., Inc., 1974.

Boylan, Richard J. and Lee K. Boylan. *Close Extraterrestrial Encounters: Positive Experiences with Mysterious Visitors*. Newberg, OR: Wild Flower Press, 1994.

Brown, Tom, Jr. *The Vision*. New York: The Berkley Publishing Corp, 1988.

Bryan, C. D. B. *Close Encounters of the Fourth Kind: Alien Abduction, UFOs and the Conference at M. I. T.* New York: Alfred A. Knopf, Inc., 1995.

Bryant, Alice and Linda Seebach. *Healing Shattered Reality*. Newberg, OR: Wild Flower Press, 1991.

Buhler, Walter K., Guilherme Pereira and Ney Matiel Pires. *UFO Abduction at Mirassol: A Biogenetic Experiment*. Tucson, AZ: Wendelle C. Stevens, 1985.

Casellato, Rodolfo R., Joao Valerio da Silva and Wendelle C. Stevens. *UFO Abduction at Botucatu: A Preliminary Report*. Tuscon, AZ: UFO Photo Archives, 1985.

Collings, Beth and Anna Jamerson. *Connections: Solving Our Alien Abduction Mystery*. Newberg, OR: Wild Flower Press, 1996.

Crystall, Ellen. *Silent Invasion: The Shocking Discoveries of a UFO Researcher*. New York: Paragon House, 1991.

Druffel, Ann and D. Scott Rogo. *The Tujunga Canyon Contacts: Updated Edition*. New York: New American Library, 1980, 1988.

Eaton, Evelyn. *The Shaman and the Medicine Wheel*. Wheaton, IL: Theosophical Publishing House, 1982.

Fearheiley, Don. *Angels Among Us*. New York: Avon Books, 1993.

Fiore, Edith. *Encounters: A Psychologist Reveals Case Studies of Abductions By Extraterrestrials*. New York: Doubleday, 1989.

Fowler, Raymond E. *The Andreasson Affair*. Newberg, OR: Wild Flower Press, 1994 Reprint.

----. *The Andreasson Affair, Phase Two: The Continuing Investigation of a Woman's Abduction by Extraterrestrials*. Newberg, OR: Wild Flower Press, 1994 Reprint.

----. *The Allagash Abductions*. Newberg, OR: Wild Flower Press, 1993.

----. *The Watchers*. New York: Bantam, 1990.

----. *The Watchers II: Exploring UFOs and the Near-Death Experience*. Newberg, OR: Wild Flower Press, 1995.

----, ed. *MUFON Field Investigator's Manual*. Seguin, TX: Mutual UFO Network, Inc., 1983.

Fuller, John G. *The Interrupted Journey*. New York: Dell Publishing, Inc., 1966.

Good, Timothy. *Above Top Secret: The Worldwide UFO Cover-up*. New York: William Morrow & Co., Inc., 1988.

----. *Alien Contact: Top-Secret UFO Files Revealed*. New York: William Morrow & Co., Inc., 1993.

----. *Alien Update*. London: Arrow Random House, 1993.

Green, Gabriel and Warren Smith. *Let's Face the Facts about Flying Saucers*. New York: Popular Library, 1967.

Haley, Leah. *Lost Was The Key*. Tuscaloosa, AL: Greenleaf Publications, 1993.

Hall, Richard. *Uninvited Guests: A Documented History of UFO Sightings, Alien Encounters and Coverups*. Santa Fe, NM: Aurora Press, 1988.

Hickson, Charles and William Mendez. *UFO Contact at Pascagoula*. Tucson, AZ: Wendelle C. Stevens, 1983.

Holzer, Hans. *The Ufonauts: New Facts on Extraterrestrial Landings*. Greenwich, CT: Fawcett Publications, Inc. 1976.

Hopkins, Budd. *Intruders: The Incredible Visitations at Copley Woods*. New York: Random House, 1987.

----. *Missing Time: A Documented Study of UFO Abductions.* New York: Richard Marek Publishers, 1981.

Jacobs, David M. *Secret Life: Firsthand Accounts oF UFO Abductions.* New York: Simon & Schuster, 1992.

Kannenberg, Ida. *The Alien Book of Truth.* Newberg, OR: Wild Flower Press, 1993.

----. *UFOs and the Psychic Facter.* Newberg, OR: Wild Flower Press, 1992.

----. *Project Earth.* Newberg, OR: Wild Flower Press, 1995.

Knight, J. Z. *A State of Mind: My Story.* New York: Warner Books, 1987.

LaVigne, Michelle. *The Alien Abduction Survival Guide.* Newberg, OR: Wild Flower Press, 1995.

Lindemann, Michael. *UFOs and the Alien Presence.* Newberg, OR: Wild Flower Press, 1995 Reprint.

Lorenzen, Coral and Jim Lorenzen. *Abducted! Confrontations with Beings From Outer Space.* New York: Berkley Publishing Corp., 1977.

Mack, John E. *Abduction: Human Encounters with Aliens.* New York: Charles Scribner's Sons, 1994.

Menger, Howard. *From Outer Space To You.* Clarksburg, WV: Saucerian Books, 1959.

O'Brien, Christopher. *The Mysterious Valley.* New York: St. Martin's Press, 1996.

Pallman, Ludwig F. and Wendelle C. Stevens. *UFO Contact From Planet Itibi-Ra.* Tucson, AZ: UFO Photo Archives, 1986.

Randazzo, Joseph. *The Contactees Manuscript.* Studio City, CA: The UFO Library Limited, 1993.

Randle, Kevin D. *The October Scenario: UFO Abductions, Theories About Them and a Prediction of When They Will Return.* Iowa City, IA: Middle Coast Publishing, 1988.

Randles, Jenny. *Abduction: Over 200 Documented UFO Kidnappings Exhaustively Investigated.* Clerkenwell Green, London: Robert Hale Limited, 1988.

Ring, Kenneth. *The Omega Project.* New York: William Morrow, 1992.

Rogo, D. Scott. *The Haunted Universe: A Psychic Look at Miracles, UFOs and Mysteries of Nature.* New York: New American Library, Inc., 1977.

Sanchez-Ocejo, Virgilio and Wendelle C. Stevens. *UFO Contact From Undersea: A Report of the Investigation.* Tucson, AZ: Wendelle C. Stevens, 1982.

Smith, Warren. *The Book of Encounters*. New York: Kensington Publishing Corp, 1976.
Steiger, Brad and Joan Writenhour. *Flying Saucers are Hostile*. New York: Award Books, 1967.
Steiger, Brad. *The Other*. New Brunswick, NJ: Inner Light Publications, 1992.
Steiger, Brad and Sherry Hansen Steiger. *The Rainbow Conspiracy*. New York: Windsor Publishing Corp, 1994.
Steiger, Brad. *The UFO Abductors*. New York: Berkley Books, 1988.
Stevens, Wendelle C. *UFO Contact From The Pleiades: A Preliminary Investigative Report*. Tucson, AZ: Wendelle C. Stevens, 1982.
----. *UFO Contact From The Pleiades: A Supplementary Investigation*. Tucson, AZ: UFO Photo Archives, 1989.
----. *Message From The Pleiades: The Contact Notes of Eduard Billy Meier, Vol. 2*. Tucson, AZ: UFO Photo Archives, 1990.
Stranges, Frank E. *My Friend From Beyond Earth*. PO Box 5, Van Nuys, CA: I. E. C. Inc., 1981.
Strieber, Whitley. *Breakthrough: The Next Step*. New York: William Morrow & Co., Inc., 1995.
----. *Communion*. New York: William Morrow & Co., Inc., 1987.
----. *Transformation: The Breakthrough*. New York: Beech Tree Books/ William Morrow & Co., Inc., 1988.
Stringfield, Leonard. *Situation Red: The UFO Siege*. New York: Fawcett Crest Books, 1977.
Teets, Bob. *West Virginia UFOs: Close Encounters in the Mountain State*. Terra Alta, WV: Headline Books, Inc., 1995.
Thompson, Richard. *Alien Identities: Ancient Insights Into Modern UFO Phenomena*. San Diego, CA: Govardhan Hill Publishing, 1993.
Turner, Karla. *Into the Fringe: A True Story of Alien Abduction*. New York: Berkley Books, 1992.
Twiggs, Denise Rieb and Bert Twiggs. *Secret Vows: Our Lives with Extraterrestrials*. New York: Berkley Books, 1995.
Vallée, Jacques. *Confrontations: A Scientist's Search for Alien Contact*. New York: Ballantine Books, 1990.
----. *The Invisible College: What a Group of Scientists Has Discovered About UFO Influences on the Human Race*. New York: E. P. Dutton, 1975.
Walton, Travis. *The Walton Experience*. New York: Berkley Publishing Corp., 1978.

Wilson, Katharina. *The Alien Jigsaw*. Portland, OR: Puzzle Publishing, 1993.

**Magazines, Journals and Newspapers:**

Adams, John. "To Whom It May Concern," *Missing Link*, Federal Way, WA: UFO Contact Center International, No. 119, (Aug. 1992), pp. 11-13.

Beckley, Timothy Green. "Dean Anderson's Ten-Year Contact Saga," *UFO Universe*, New York: Condor Books, Inc., Vol. 1, No. 1, (July 1988), p. 7.

Chorvinsky, Mark. "Our Strange World," *Fate*, St. Paul, MN: Llewellyn Worldwide, Ltd., (Oct. 1993), pp. 22-28.

Coats, Rusty. "Believers Share Stories," *Bee*, Modesto, CA, (Jan 9, 1994). (see also: *UFO Newsclipping Service*, Route 1, Box 220, Plumerville, AR 72127, (Feb. 1994), No. 295, pp. 8-9).

Condon, Christopher. "World, Other Planets, Represented at UFO Congress in Budapest," *Budapest Week*, Budapest, Hungary, (Oct. 29—Nov. 4, 1992). (see also: *UFO Newsclipping Service*, (Feb. 1993), No. 283. p. 13).

Culver, C. Leigh. "An Unforgettable Close Encounter," *UFO Encounters*, Norcross, GA: Aztec Publishing, Vol. 2, No. 3, pp. 3-7, 29-30.

Dennett, Preston. "Contactee: Firsthand," *UFO*, Los Angeles, CA, California UFO. Vol. 5, No. 5, (Sep./Oct. 1990), pp. 36-39.

Dongo, Paul. "The Paranormal: A Wide Wonderful World," *Missing Link*, Texas City, TX: Living Light Productions, No. 131, (March/April 1994), p. 24.

Haggerty, Alice. "Abductee, USA," *International UFO Library Magazine*, 11684 Ventura Blvd, Studio City, CA: UFO Library, Inc., Vol. 2, (1991), p. 34.

Hamilton, Bill. "A Case of Medical Intervention," *MUFON UFO Journal*, Seguin, Texas: The Mutual UFO Network, No. 328, (Aug. 1995), pp. 14-15.

Huneeus, Antonio. "Close Encounters in Peru," *UFO Universe*, New York: GCR Publishing Group, Inc., Vol. 4, No. 1, (Spring 1994), pp. 52-54.

Johnson, Paula. "Can Ufonauts Cure Us of Aids, Cancer and Other Fatal Diseases?" *UFO Universe*, New York: Condor Books, Inc., Vol. 1, No. 1, (July 1988), pp. 24-27.

Leupold, Edwin H. "Deming Woman Recounts 50 Years of Real-life UFO 'Close Encounters,'" *Headlight*, Deming, NM, (Jan. 27, 1994). (see also: *UFO Newsclipping Service*, No. 297, (April 1994), p. 4).

Mesnard, Joel. (translated by Claudia Yapp) "The French Abduction File," *MUFON UFO Journal*, Seguin, TX: Mutual UFO Network, Inc., No. 309, (Jan. 1994), pp. 10-11.

Morrow, Helga. "The Sedona Chronicles," *Missing Link*, Federal Way, WA: UFO Contact Center International, No. 124, (Jan./Feb. 1993), pp. 15-17.

Morton, Susan Nevarez. "They're Here—First Place," *Express News*, San Antonio, TX, (Feb. 26, 1989). (see also: *UFO Newsclipping Service*, No. 237, (April 1989), p. 11).

Norris, Michael. "The First Annual Gulf Breeze UFO Conference: Part II," *UFO Encounters*, Norcross, GA: Aztec Publishing, Vol. 1, No. 12, p. 7.

Overstreet, Sarah. "Search of Space Conjures Up Spooky Stories," *News Record*, Miami, OK, (Oct. 27, 1992). (see also: *UFO Newsclipping Service*, No. 281, (Dec. 1992), p. 11).

Paxton, Julie. "Struck by Lightning," *Fate*, St. Paul, MN, Vol. 48, No. 4, Issue #541, (April 1995), pp. 79-80.

Robinson, Malcolm. "The International Page," *Missing Link*, No. 126, (May/June 1993), pp. 18-19.

Sable, Patricia. "The Light From Within: The Strange Story of a UFO Abductee," *UFO Universe*, New York: Charlotte Magazine Corp., Vol. 1, No. 1, (Feb./March 1991), pp. 14-18, 66.

Salter, Jr., John. "No Intelligent Life Is Alien To Me," *Contact Forum*, Newberg, OR: Wild Flower Press, Vol. 1, No. 1, pp. 1-3.

Schmidt, Steve. "Friendly Aliens?" *Herald*, Grand Rapids, ND, (Nov. 11, 1989). (see also: *UFO Newsclipping Service*, No. 245, (Dec. 1989), p. 6).

Steiger, Brad and Sherry Hansen Steiger. "UFOS: Friend Or Foe?" *UFO Universe*, New York: Charlotte Magazine Corp., Vol. 1, No. 5, (Oct./Nov. 1991), p. 43.

Tessman, Diane. "Three Amazing European Close Encounters," *UFO Universe*, Vol. 2, No. 4, (Winter 1993), pp. 62-64.

Townsend, Peggy R. "Close Encounters of the Watsonville Kind," *Sentinel*, Santa Clara, CA, (Nov. 11, 1993). (see also: *UFO Newsclipping Service*, No. 294, (Jan. 1994), p. 1).

Traveler, Star. "'Clown Healers' From Reticulum," *Contact Forum*, Newberg, OR: Wild Flower Press, Vol. 3, No. 2, (March/April 1995), pp. 9-10.

Turner, Karla. "Aliens—Friends or Foes?" *UFO Universe*, New York: Charlotte Magazine Corp., Vol. 3, No. 1, p. 12.

Wright, Dan. "The Entities: Initial Findings of the Abduction Transcription Project—A MUFON Special Report, Part Two," *MUFON UFO Journal*, Seguin, TX: Mutual UFO Network, No. 311, (March 1994), pp. 3-6.

**Radio/Television Programs:**

*The Other Side*. NBC.

*The Oprah Winfrey Show*. NBC.

*UFOs Tonite with Don Ecker*. The Cable Radio Network (CRN) on public access television. Saturday evening 9:00-11:00 PM. Based in Sunland, California.

*Visitors From the Unknown*. Writer, Michael Grais; Producer, Sharon Gayle; Director, Penelope Spheeris.

# Index


A

*A State of Mind* 176
*Abducted!* 176
*Abduction* (Mack) xiii, 85, 176
*Abduction* (Randles) 176
*Above Top Secret* 175
Adams, John 103, 155, 178
Adler, Olga 48, 151
AIDS xi, xiii, 114, 115, 166
*Alien Book of Truth, The* 176
*Alien Contact* 175
*Alien Identities* xv, 177
*Alien Jigsaw* 98, 178
*Alien Update* 175
*Allagash Abductions, The* 175
Anderson, Dean 95
*Andreasson Affair, Phase Two, The* 175
*Andreasson Affair, The* 175
Andreasson, Betty 3, 56
aneurysm 90, 152, 158
angels 96, 143, 144
*Angels Among Us* 175
angioma 92, 154, 158
ankle 19
Appleton, Cynthia 33, 129
arthritis 97, 153, 158, 159
asthma 81, 145, 153, 158

B

B., Denise 121
back 29, 41
    back pain 48, 49, 51, 158
Bartsch, Leo 20, 150
Beckley, Timothy Green xvi, 19, 20, 95, 118, 178
*Bee* (newspaper) 178
Bell, Fred 22, 151, 174
Bershad, Michael 4, 134
*Beyond Earth* 174
Bishop, Denise 12
black death 89
Blackman, Brandon 19, 151
blindness 58, 142, 143, 149, 158
Blum, Judy xiv, 56, 66, 97, 133, 174
Blum, Ralph xiv, 56, 66, 97, 133, 174
Boas, Antonio Villas 11
Bochereshoni, Professor 135
body 31
    body injuries 158
*Book of Encounters, The* 177
Bowen, Charles 12, 174
Bowles, Joyce 135
Boyd, Doug 140, 174
Boylan, Lee K. 174
Boylan, Richard J. xvi–xvii, 128, 174
*Breakthrough* 166, 177
Brown, Jr., Tom 140, 141, 174
Bryan, C. D. B. 57, 115, 174
Bryant, Alice 174
bubonic plague 89
*Budapest Week* (newspaper) 178
Buhler, Walter K. 128, 174
Bullard, Thomas E. xiv
burns 22, 27, 34, 146, 150, 151, 154, 158
bursitis 51, 156, 158

C

Caesarean section 29, 158
cancer xi, xvii, 34, 79, 100, 117–130, 134, 143, 145, 150, 151, 155, 156, 158
    bowels 119
    breast 158
    colon 153, 158
    hip 118
    lung 125, 155, 158
    pancreas 118
    skin 118, 124, 130, 150, 152, 158
    stomach 120, 121, 126, 155, 158
    throat 121, 151, 158
candidiasis 158
Cardenas, Filiberto 12
Carlos, Edward 85, 99, 124, 149, 152, 154, 160
Carlsburg, Kim 136
Carpenter, John 57, 163

Carvalho, Dr. 13
Case #
#001 94, 149
#002 85, 149
#003 85, 149
#004 58, 149
#005 100, 149
#006 30, 149
#007 36, 149
#008 47, 150
#009 61, 150
#010 129, 150
#011 125, 150
#012 33, 150
#013 37, 150
#014 20, 150
#015 25, 150
#016 96, 150
#017 17, 150
#018 82, 150
#019 77, 150
#020 73, 150
#021 19, 150
#022 56, 150
#023 97, 151
#024 30, 151
#025 45, 151
#026 31, 151
#027 22, 151
#028 106, 151
#029 66, 151
#030 48, 151
#031 114, 151
#032 121, 151
#033 21, 151
#034 19, 151
#035 41, 151
#036 118, 151
#037 33, 151
#038 76, 151
#039 83, 152
#040 22, 152
#041 84, 152
#042 89, 152
#043 95, 152
#044 24, 152
#045 75, 152
#046 118, 152
#047 105, 152
#048 29, 152
#049 68, 152
#050 90, 152
#051 112, 152
#052 124, 152
#053 31, 152
#054 68, 153
#055 46, 153
#056 27, 153
#057 97, 153
#058 111, 153
#059 42, 153
#060 98, 153
#061 67, 153
#062 28, 153
#063 81, 153
#064 57, 153
#065 98, 153
#066 153
#067 49, 122, 153
#068 99, 154
#069 78, 154
#070 49, 154
#071 109, 154
#072 34, 154
#073 60, 154
#074 60, 154
#075 108, 154
#076 154
#077 112, 154
#078 108, 154
#079 26, 154
#080 92, 154
#081 104, 154
#082 107, 155
#083 103, 155
#084 125, 155
#085 120, 155
#086 104, 155
#087 101, 155
#088 50, 155
#089 57, 155
#090 25, 155
#091 113, 155
#092 124, 155
#093 37, 155
#094 102, 155
#095 102, 155
#096 51, 155
#097 43, 123, 155
#098 156
#099 79, 156
#100 44, 156
#101 50, 156
#102 50, 156

#103 51, 156
#104 51, 156
#105 113, 156
Casellato, Rodolfo R. 91, 92, 174
Cash, Betty 10
chest 28
Chi Gong 136
cholera 89
Chorvinsky, Mark 48, 178
Clamser, Mary 146
*Close Encounters of the Fourth Kind* 174
*Close Extraterrestrial Encounters* xvi, 174
Coats, Rusty 163, 178
cold 43, 51, 69, 153, 158
colitis 113, 155, 158
color blindness 57, 155, 158
*Communion* 177
Compardo, Kathleen 128
Condon, Christopher 125, 178
*Confrontations* 177
*Connections* 174
*Contact Forum* (newsletter) 179, 180
*Contactees Manuscript, The* 176
Cooper, Dr. 4
Creighton, Gordon 125, 133
Crystall, Ellen 47, 150, 175
Culver, C. Leigh 113, 178
Cyr, Jean 89, 152
cyst xvi
D
D., Daniel 36, 154
da Silva, Joao Valerio 174
Davidson, Licia 122–123, 153
Davis, Charles Keith 13
DeGroot, Jan 67, 153
Delmundo, David 75
Dennett, Preston 178
Desmond, Sammy 3
DeSoto, Anne 97, 153
diabetes 102, 155, 158
diphtheria 89, 95, 96, 150, 158
dizziness 49, 153, 158
Dongo, Paul 178
Doyle, Chuck 41, 151
Druffel, Ann 2, 3, 130, 175
E
ear 41, 149
ear infection 30, 158
ear surgery 166
earache 167
Eaton, Evelyn 141, 175
*Encounters* 99, 175
*Express News* (newspaper) 179
eyes 60, 61, 154, 158, 159
F
*Fate* (magazine) 120, 178, 179
Fearheiley, Don 144, 175
feet 25
fever 41, 45, 74, 151, 158
Filho, Joao Prestes 13
finger 17, 20, 30
Fiore, Edith xii, 3, 24, 44, 50, 51, 79, 92, 93, 113, 124, 134, 175
flesh wound 17–38, 149, 153, 158
ankle 150, 158
back 151, 153
back pain 154, 155
body 150, 151, 152
Caesarean section 152
finger 150, 151, 158
foot 154
genitals 158
hand 152, 158
head injury 155
knee 155
leg 143, 153
legs 158
rib 151
*Flying Saucer Review* (magazine) 125
*Flying Saucers are Hostile* 177
Flynn, James 13
Fontes, Olavo T. 125
foot 35
Fowler, Raymond E. 3, 10, 145, 175
*From Outer Space To You* 176
Fuller, John G. 4, 175
G
Garreau, Gilles 121
Gayle, Sharon 106, 180
genital 36
Godfrey, Alan 105, 152
Good, Timothy 11, 12, 26, 175
Goode, Robert W. 17, 150
Grais, Michael 106, 180
Green, Gabriel 18, 175
Grevler, Ann 25, 150
H
Haggerty, Alice 96, 150, 178
Haley, Leah 56, 175
Hall, Richard 13, 14, 118, 175
Hamilton, Bill 178
Hamilton, William 108, 109
hand 21
Harder, James 114
*Haunted Universe, The* 176
head 22, 24, 25
head cold 42, 151
head injury 158
head wound 152
*Headlight* (newspaper) 179
*Healing Shattered Reality* 174

heart xvi, 98, 99, 100, 101, 102, 149, 153, 154, 155, 158
heart disease xi
Hendricks, Beryl 145
*Herald* (newspaper) 179
Hessemann, Michael 101
Hickson, Charles 175
Higdon, Carl 76, 83, 151, 152, 160
Hill, Barney 3
Hill, Betty 4
hip 51, 156, 158
Hodgkins disease 166
Holzer, Hans 6, 128, 175
Hopkins, Budd xiii, 4, 5, 12, 27, 60, 108, 134, 175
Horton, Virginia 12
*Humanoids, The* 174
Huneeus, Antonio xv, 120, 121, 178
I
infertility 104, 105, 106, 151, 152, 154, 159
integumentary system 65—70, 153, 158, 159
*International UFO Library* (magazine) 96, 178
*Interrupted Journey, The* 175
intestinal 50, 51, 113, 150, 154, 156, 159
*Into the Fringe* 177
*Intruders* xiii, 27, 108, 175
*Invisible College, The* 177
J
Jacobs, David M. xii, xiii, 5, 85, 96, 176
Jesus Christ xv, 143
Johnson, Paula xvii, 21, 83, 120, 178
Jordan, Debbie 27, 108, 154
Julia (of *Abduction*) 135
K
Kannenberg, Ida 176
kidney 73, 158
disease 74, 150, 159
failure 159
stones 75, 76, 77, 83, 151, 152, 159
Kilburn, Steve 4
knee 37, 159
Knight, J. Z. 141, 142, 176
Kurz, Shane 6
L
Lamb, Barbara 35, 49, 100, 101, 106, 136
Landrum, Colby 10
Landrum, Vickie 10
Lauritzen, Hans 77, 150
Leach, L. R. 18
legs 26, 46, 150
leprosy 143
*Let's Face the Facts about Flying Saucers* 175
leukemia xiii
Leupold, Edwin H. 50, 179
liver 73, 150, 154, 156, 158, 159
liver disease xi, 78, 79
liver hepatitis 77
Lorenzen, Coral 11, 114, 176
Lorenzen, Jim 11, 114, 176
*Lost Was The Key* 56, 175
Luca, Bob 3
lung 81, 82, 86, 150, 158, 159
Lupus 167
M
Maceiras, Ventura 66, 151
Mack, John E. xiii, 35, 85, 86, 99, 100, 124, 133, 135, 176
Maria, Anazia 126
Martins, Joao 125
McCoy, Billy 17
medical injuries from UFOs 9—14
Meier, Eduard 32—33, 84, 85, 149, 151, 152, 160
Mendez, William 175
Menger, Howard 58, 59, 149, 176
Menninger Foundation 139
Mesnard, Joel 122, 179
*Message From The Pleiades* 177
Michalak, Stephen 11
missing fetus syndrome 1, 35, 104, 110
*Missing Link* (magazine) 178, 179
*Missing Time* xiii, 4, 12, 134, 176
moles 67, 68, 152, 159
Morgan, Karen 5
Morrow, Helga 37, 51, 102, 135, 179
Morton, Susan Nevarez 37, 179
*MUFON Field Investigator's Manual* 175
*MUFON UFO Journal* 178, 179, 180
multiple sclerosis 146, 152, 159
muscular dystrophy 94, 95, 152, 159
Mutual UFO Network (MUFON) xiii, 9, 10, 26, 27, 134
*My Friend From Beyond Earth* 177
myopia 56, 97, 150, 159
N
near-death experience 31, 32, 145, 146
*News Record* (newspaper) 179
Norris, Michael 163, 179
numbness 20, 150, 159
O
O'Brien, Christopher 176
*October Scenario, The* 176
*Omega Project, The* 176
One 9
*Oprah Winfrey Show, The* (television show) 180
*Other Side, The* (television show) 180
*Other, The* 177
Overstreet, Sarah 94, 179
P
Pallman, Ludwig F. 73—75, 150, 176
Pam (of *Abduction*) 135

paralysis 19, 112, 150, 152, 159
Paul (of *Abduction*) 135
Paxton, Julie 146, 179
Pereira, Guilherme 174
Pires, Ney Matiel 174
pleurisy 25
pneumonia xi, xiii, 83, 85, 143, 149, 152, 159
polio 89, 103, 104, 140, 155, 159
poliomyelitis xiii
Ponce de Leon, Anton 120
*Project Earth* 176
*Promise, The* 174
pseudonym
  Kilburn, Steve 134
pseudonyms
  Barbara, Benara 167
  Champlain, Anthony 22, 152
  Collings, Beth 60, 61, 154, 174
  Davis, Kathie (*See* Jordan, Debbie) 27, 108, 154
  Hansen, Alicia 31, 109, 137, 152, 154, 160
  Jamerson, Anna 60, 61, 154, 174
  Michelline 34, 154
  Miller, Lynn 96
  Shaw, Sara 2, 129
  Ted 92
  X., Ann 50, 155
  X., Barbara 3
  X., Betty 46
  X., David 154
  X., Doctor 19, 150
  X., Doriel 104, 154
  X., Eddie 155
  X., Elizabeth 124
  X., Fred 27
  X., Helen 118, 151
  X., James 24, 152
  X., Julinho 126
  X., Lais 126
  X., Linda 43, 44, 50, 51, 79, 113, 123, 134, 155, 156, 160
  X., Mae 98, 153
  X., Michael 118, 152
  X., Mr. 120, 155
  X., Otavinho 127
  X., Sherry 44, 156
  X., Ted 154
  X., Tom 3

R

R., Carl 68
R., Dagmar 68, 152
*Rainbow Conspiracy, The* 177
Ramtha 141
Randazzo, Joseph 30, 31, 46, 47, 104, 111, 176
Randle, Kevin 114, 176
Randles, Jenny 34, 129, 135, 176
Rayna 163
Reiki 136
rheumatism 56, 97, 151, 159
ribs 33, 159
Ring, Kenneth 145, 146, 176
Roach, Deborah 114, 151
Roach, Pat 114
Robinson, Diane 144
Robinson, Kelly 5–6
Robinson, Malcolm 108, 179
Rogo, D. Scott xv, 130, 175, 176
*Rolling Thunder* 174
Romaniuk, Pedro 66
Rudder, Pierre de 142
Rylka, Richard 30, 31, 104, 111, 115, 149, 151, 153, 155, 160

S

Sable, Patricia 62, 179
Salter, Jr., John 68, 69, 153, 179
Sanchez-Ocejo, Virgilio 12, 76, 176
Schmidt, Steve 69, 179
Scott (of *Abduction*) 135
*Secret Life* 5, 85, 96, 176
*Secret Vows* 177
Seebach, Linda 174
*Sentinel* (newspaper) 179
*Shaman and the Medicine Wheel, The* 175
Shenefield, Marianne C. 61, 150
*Silent Invasion* 47, 175
sinus xvi, 50, 155, 159, 167
*Situation Red* 177
small pox 89
Smith, Louise 11
Smith, Warren 175, 177
Smith, Yvonne 114
Spheeris, Penelope 106, 180
Stafford, Mona 11
Stargarrdt's disease 61
Steiger, Brad xiv, 13, 50, 68, 77, 78, 79, 83, 98, 104, 105, 112, 177, 179
Steiger, Sherry 50, 79, 98, 112, 177, 179
Stevens, Wendelle 33, 85, 90, 125, 126, 174, 176, 177
stomach 123, 159
*Strange Magazine* (magazine) 48
Stranges, Frank E. 21, 151, 177
Strieber, Whitley 166, 177
Stringfield, Leonard xiv, 42, 66, 177
Sturdevant, Harry 13
sty 57, 153, 159

T
T., Richard 112, 152
Tai Chi 136
teeth xiv, 36, 66, 154, 159
Teets, Bob 36, 37, 177
Tessman, Diane 67, 179
*The Mysterious Valley* 176
*The Omega Project* 145
*The Other Side* (television program) 136
Thomas, Elaine 11
Thompson, Richard xv, 177
Thor, Valiant 21
Thunder, Rolling 139
Tomey, Debbie 108
Townsend, Peggy R. 97, 179
*Transformation* 177
Traveler, Star 24–25, 155, 180
tuberculosis 83, 93, 149, 152, 159
*Tujunga Canyon Contacts, The* 2, 175
tumor 107, 108, 109, 111, 113, 140, 142, 145, 153, 154, 155, 156, 159
Turner, Karla 28, 56, 57, 177, 180
Twiggs, Bert 28, 29, 42, 43, 57, 81, 82, 113, 124, 125, 153, 155, 160, 177
Twiggs, Christopher 57, 153
Twiggs, Denise 28, 29, 42, 43, 57, 81, 82, 113, 124, 125, 152, 155, 160, 177
Twiggs, Stacey 153
typhoid 89
U
*UFO* (magazine) 178
*UFO Abduction at Botucatu* 91, 92, 174
*UFO Abduction At Mirassol* 126
*UFO Abduction at Mirassol* 174
*UFO Abductions* xiv
*UFO Abductors, The* 177
*UFO Contact at Pascagoula* 175
*UFO Contact From Planet Itibi-Ra* 176
*UFO Contact From The Pleiades* 177
*UFO Contact From Undersea* 176
*UFO Encounters* (magazine) 178, 179
*UFO Newsclipping Service* (newspaper) 178, 179
UFO Photo Archives 91, 126
*UFO Universe* (magazine) xvi, 120, 178, 179, 180
*Ufonauts, The* 175
*UFOs and the Psychic Facter* 176
*UFOs Tonite with Don Ecker* (television program) 108, 180
ulcer 50, 156, 159
*Uninvited Guests* 13, 175
uterus 128
V
Valerio, Joao 91
Vallée, Jacques xvi, 13, 19, 142, 143, 177
Van Hoestenberghe, Dr. 142
Van Klausen, Morgana 108
Van Klausens, Morgana 154
Vasquez, Hector 75, 152
Vicente, Joao 90, 152
Virgin Mary 142, 143
*Vision, The* 174
*Visitors From the Unknown* (television program) 106, 180
W
Walker, Ralph 143
*Walton Experience, The* 177
Walton, Travis 5, 177
warts 41, 67, 159
*Watchers II, The* 175
*Watchers, The* 175
Wells, Sixto Paz 46
*West Virginia UFOs* 36, 177
White, Fred 82, 150
Whitley, Jan 129, 136
Wills, Jerry 45, 151
Wilson, Katharina 98, 99, 153, 178
Wolf, Stalking 140
Wright, Dan xiii, xiv, 180
Writenhour, Joan 177
Y
Yapp, Claudia 179
yeast infection xvi, 43, 44, 51, 128, 155, 156, 158

Additional books offered by Blue Water Publishing, Inc.

## Wild Flower Press

**CONNECTIONS**
*Solving Our Alien Abduction Mystery*
Beth Collings & Anna Jamerson

**THE WATCHERS II**
*Exploring UFOs and the Near-Death Experience*
Raymond E. Fowler

**THE ALIEN ABDUCTION SURVIVAL GUIDE**
*Practical Advice and Solutions*
Michelle LaVigne

**UFOs and THE ALIEN PRESENCE**
*Six viewpoints by Stanton Friedman, Linda Moulton-Howe, Budd Hopkins and others*
Editor: Michael Lindemann

**PROJECT EARTH**
*From the ET Perspective*
Ida M. Kannenberg

**BECOMING GODS**
*Prophecies, Spiritual Guidance and Practical Advice for the Coming Changes*
by Cazekiel as received by James Gilliland

**THE ALLAGASH ABDUCTIONS**
*Undeniable Evidence of Alien Intervention*
Raymond E. Fowler

**BIGFOOT MEMOIRS**
*My Life with the Sasquatch*
Stan Johnson

*For ordering information, see next page...*

**Swan•Raven & Co.**

**A MAGICAL UNIVERSE**
The Best of Magical Blend Magazine
Michael Langevin and Jerry Snider

**PLANT SPIRIT MEDICINE**
*Healing with the Power of Plants*
Eliot Cowan

**TAROT OF THE SOUL**
*A guiding oracle that uses ordinary playing cards*
Belinda Atkinson

**CALLING THE CIRCLE**
*The First and Future Culture*
Christina Baldwin

**RITUAL**
*Power, Healing and Community*
Malidoma Somé

**WHEN SLEEPING BEAUTY WAKES UP**
*A Woman's Tale of Healing the Immune System and Awakening the Feminine*
Patt Lind-Kyle